Basic Mathematics for Students of Air Pollutants

Basic Mathematics for Students of Air Pollutants

By

Robert Maynard

University of Birmingham, UK
Email: robertmaynard3@gmail.com

and

Richard Atkinson

St George's, University of London, UK
Email: atkinson@sgul.ac.uk

ROYAL SOCIETY
OF **CHEMISTRY**

Print ISBN: 978-1-83767-278-3
PDF ISBN: 978-1-83767-279-0
EPUB ISBN: 978-1-83767-280-6

A catalogue record for this book is available from the British Library

The Royal Society of Chemistry is a charity, registered in England and Wales, Number 207890, and a company incorporated in England by Royal Charter (Registered No. RC000524), registered office: Burlington House, Piccadilly, London W1J 0BA, UK, Telephone: +44 (0) 20 7437 8656.

Visit our website at books.rsc.org

Printed in the United Kingdom by CPI Group (UK) Ltd, Croydon, CR0 4YY, UK

Preface

The aim of this short book is to introduce the reader to the basic
mathematics needed for study of the effects of air pollutants on
health. Our belief that such a book might be helpful has grown out of
our experience of teaching, which has shown that although students
may have studied much of the necessary material in their school-days,
their knowledge may have become a little rusty and some aspects of
basic mathematics are poorly understood. We have also found that
students find difficulty in reading more advanced books, for example in the area of aerosol science, because the authors of such works
assume that their readers are familiar with higher mathematics. For
some students this may be true, for others it is not.

How much mathematics do students need to know? They certainly
need to know something of elementary algebra, of logarithms and of
a few statistical distributions. The student also needs to know something of elementary descriptive statistics. Another area which requires
some mathematical background is met as soon as the techniques
of statistical regression are encountered. These, then, are the main
topics discussed in the more mathematical chapters of this book. In
addition, we have said a little about the mathematical description of
aerosols. Much of what is now known about the effects of air pollutants on health has been derived by epidemiological studies; we have
included a chapter outlining the methods most frequently used. This
has led us to the difficult question of causality, of deciding whether
statistically significant associations are causal in nature. Lastly, we
have discussed how the effects of air pollutants on health may be
quantified.

Basic Mathematics for Students of Air Pollutants
By Robert Maynard and Richard Atkinson
© Robert Maynard and Richard Atkinson 2024
Published by the Royal Society of Chemistry, www.rsc.org

It is possible that having read this introductory book the reader will be encouraged to look further into the subject and, to help with this, we have provided a Further Reading section of books we have found useful. The selection of books is entirely personal; no doubt there are other equally good, perhaps better, books. No textbooks of higher mathematics have been included. One of us (RM) wishes to acknowledge, now, the debt he owes to Alan and Angela Crowe. Their "Mathematics for Biologists" has been his constant companion for many years, and we have adopted much of their terminology and approach in this book. We also wish to acknowledge the debt we owe to the late William Hinds whose book "Aerosol Technology" has provided much of the material used in our chapter on aerosols. The illustrative material used in this book has been borrowed from these works and from others listed in the Further Reading. We are grateful to the publishers of all these works and others for permission to use their material.

We have done our best to weed out errors from this book but some slips of the pen may remain; we would be grateful if readers would point these out to us.

Finally, we should like to thank our wives for their patience while we have been working on this book and our publishers, especially Helen Armes, for their help and encouragement.

Robert Maynard
Richard Atkinson

Contents

Basic Mathematics for Students of Air Pollutants
By Robert Maynard and Richard Atkinson
© Robert Maynard and Richard Atkinson 2024
Published by the Royal Society of Chemistry, www.rsc.org

3 Exponential Decay and Semi-logarithmic Graph Paper   18

4 Other Curvilinear Relationships   25

5 Preliminary Examination of a Set of Measurements   29

6 The Normal Distribution Curve   45

7 Does the Normal Distribution Curve Fit the Data? 62

8 Distribution of the Diameters of Particles in a Typical Aerosol 69

9 The Statistical Distribution of Mass and Surface Area of Particles Comprising an Aerosol 81

10 Deposition of Particles in the Respiratory Tract   85

11 Gases, Liquids and Droplets   102

12 Elementary Considerations of Line Fitting Techniques: Derivation of Concentration–Response Relationships   113

13 Air Pollution and Health 134

14 Causality 151

15 Quantification of the Effects of Air Pollutants on Health   165

Appendix   179

Further Reading   186

Subject Index   189

1 Introduction

1.1 Scope of This Book

Study of the effects of air pollutants on health requires a multi-disciplinary approach. The number of contributing disciplines may, for the beginner, appear daunting: physics, chemistry, toxicology, epidemiology, physiology, medicine, and in most of these knowledge of at least elementary mathematics and statistics is needed.

How much mathematics do we need to know? We certainly need to know something of elementary algebra, of logarithms and of a few statistical distributions. When the authors were students, logarithms were an essential aid to calculation and students used either tables of logarithms or their mechanical equivalent, the slide rule. The advent of personal computing, pocket calculators and subsequently smart phones has changed all that. Throughout the book we refer to the use of a calculator. Any basic pocket calculator or smart phone that includes simple scientific functions will suffice. Alternatively, simple software packages such as spreadsheets can also be used.

We also need to know something of elementary descriptive statistics. Most readers will have some familiarity with this area, although their knowledge may have become a little rusty; we have provided a short refresher course for such readers. As regards statistical distribution curves our main concern will be with the characteristics of such curves rather than with the equations that describe them, let alone with how those equations can be derived. We also cover how to summarise distributions using common statistics such as means, quantiles and standard deviations. We say little about hypothesis testing,

Basic Mathematics for Students of Air Pollutants
By Robert Maynard and Richard Atkinson
© Robert Maynard and Richard Atkinson 2024
Published by the Royal Society of Chemistry, www.rsc.org

instead focusing on the derivation and interpretation of confidence intervals that are commonly used in air pollution health research. In all scientific work involving the relations between one thing, for example exposure to air pollutants, and another, for example, effects on health, the techniques of statistical regression are used. The elements of linear regression are not difficult to grasp although, of course, the details of more advanced methods may be challenging, at least to the non-mathematician.

Much of the work included in this book involves the drawing of graphs. Special forms of graph paper were, until recently, widely used and are described in this book. The up-to-date readers may be amused by the thought of using graph paper and pencils and will know that such plots can be generated with the computer. Of course, they are right. There is, however, something to be said for plotting a graph by hand if only that it provides time to think about the data and what the graph means. The graphs in this book were produced using the statistical software package Stata (StataCorp., 2017. Stata Statistical Software: Release 15. College Station, TX: StataCorp LLC). There are a number of other software products that can also be used, including some freely available, *e.g.* R (R Core Team, 2021. R: A language and environment for statistical computing. R Foundation for Statistical Computing, Vienna, Austria. URL https://www.R-project.org/). Whilst these powerful software packages can certainly make life easier once one has gained some experience in using them, they still require the user to understand their data and to use the software appropriately. The need for a basic understanding of graph construction, choice of axes, and interpretation remains.

The interpretation of what a graphical representation of a relationship or a distribution is of cardinal importance. When confronted with a graph showing, for example, the distribution of particle sizes in an aerosol the reader should be able to interpret the information and to grasp what it means and should certainly be able to interpret the scales on the x and y axes. When faced with a numerical description of the characteristics of an aerosol they should be able to visualize its distribution curve. Aerosol science might be considered a sub-specialty of physics and is of great importance in the study of the effects on health of inhaled particles. Therefore, we have discussed the more elementary mathematical aspects of aerosols and the use of such descriptors as mass median diameter and count median diameter. These are, of course, explained in greater detail in textbooks of aerosol science.

One important use of air pollution data is in the study of their effects on health. The literature on this subject is vast. Much of the evidence for such effects is derived from epidemiological studies. We have, therefore, included a short chapter outlining, very briefly, some of the epidemiological techniques used in the study of the effects of air pollutants on health. Reading this chapter will not turn readers into epidemiologists but should, we hope, help them to understand research papers and reviews of epidemiological studies that investigate associations between air pollution and health.

The next and very important question is: are statistically significant associations between ambient concentrations of air pollutants and indicators of health causal in nature? Statistical analysis cannot answer this question; rather it involves considerations of toxicology in a broad sense. So important has this question of causality become in the air pollution field that we take it up in a separate chapter. Our final chapter sets out some methods used for estimating the size of the effect of air pollutants on health: quantification of the effects of air pollutants. These methods are based on epidemiological data and it should be clearly understood that the question of causality must be settled before such methods are applied.

1.2 Points of Elementary Mathematics

1.2.1 Basic Algebraic Identities

All readers will have studied the elements of algebra in their school days. Please ensure that you recognize and understand the following elementary identities. Recall that in the index notation used by mathematicians, as in a^x, a, the base, is referred to as being raised to the power of x. In the expression a^x, x is called the exponent of a and the process of "raising to the power" is called exponentiation. Note, too, that a dot is often used instead of the multiplication sign in algebraic expressions.

$$a + a = 2a$$

$$a \cdot a = a^2$$

$$-a \cdot a = -a^2$$

$$-a \cdot -a = a^2$$

$$a^x \cdot a^y = a^{x+y}$$

$$a^1 = a$$

$$a^0 = 1$$

$$\frac{a^x}{a^y} = a^{x-y}$$

$$\left(a^x\right)^y = a^{xy}$$

$$a^{-1} = \frac{1}{a}$$

$$a^{-x} = \frac{1}{a^x}$$

$$a^{1/x} = \sqrt[x]{a}$$

Here $\sqrt[x]{a}$ means the xth root of a; if $x = 2$ then $\sqrt[x]{a}$ is the square root of a (the 2 is usually omitted)

$$a^{-\frac{1}{x}} = \frac{1}{\sqrt[x]{a}}$$

$$\left(\frac{1}{x}\right)^y = \frac{1}{x^y}$$

$$\left(\frac{1}{x}a\right)^y = \frac{a^y}{x^y}$$

$$n! = n(n-1)(n-2)(n-3)........(2)(1)$$

1.2.2   Significant Figures

One is often asked to state the result of some measurement or study "to n significant figures", for example to 3 significant figures.

The standard definition (see Further Reading) of significant figures is:

Significant figures may be defined as those which must be retained for any position of the decimal point.

5567000 has 4 significant figures; whereas 200 has 1 significant figure.

Let us look at some examples which make this clear. Consider the number 12.2345678. Expressing this to 3 significant figures gives 12.2. Had the number been 12.256789 we should have rounded the first figure to the right of the decimal point: 12.3 to three significant figures.

There is no arguing with the importance of the numbers to the left of the decimal point, and we have room for just one to the right of the point if we want 3 significant figures. The reader will know that in "rounding" we examine the figure to be rounded, for example in 12.2345678 when we are rounding to three significant figures the figure to be rounded is 2 (highlighted), and the figure by which it is followed, in this example, is 3. If the immediately following figure is 5 or above we round upwards, if the immediately following figure is below 5 we round downwards.

Consider 0.0002345678. The three zeros to the right of the decimal point are NOT significant figures, they simply define where the significant figures are in relation to the point, *i.e.* the magnitude of the number. To 3 significant figures: 0.000235. This is apparent when this number is written in scientific notation, *i.e.* 2.35×10^{-4} (see next paragraph).

Consider 434000.1. To 3 significant figures: 434000, to 2 significant figures: 430000,

Consider 4040404.4. To 3 significant figures: 4040000.

Consider 4300000. To 3 significant figures: impossible. This is a trick question: there are only 2 significant figures present.

Consider 0.0002003004. To 3 significant figures: 0.000200. The zeros between the 2 and the 3 ARE significant figures, the zeros to the left of the 2 are NOT.

How do you decide the number of significant figures to use? A good question. If quoting a measurement from an instrument or technique the number of significant figures should take into account the accuracy of the device/technique, *i.e.* use the number present in the original data. It is not appropriate to report a number to 5 significant figures if the accuracy of the measurement system is only to 3 significant figures. There are also rules to help when adding/subtracting and multiplying/dividing numbers – the number of significant figures used in the answer should be the same as the least number of significant figures present in the numbers used in the calculation, *e.g.* 12.3 + 4.12 = 16.4, not 16.42.

1.2.3   E Notation

One often sees numbers on the y, ordinate, axis of a distribution of particle diameters expressed like this:

5E-7 or 5E-07 or 5E-007

These all mean the same thing: 5×10^{7} or 5 multiplied by 10 seven times or

$5 \times 10 \times 10 \times 10 \times 10 \times 10 \times 10 \times 10 = 5 \times 10,000,000 = 50,000,000$

1.3   Variates and Variables

The student may have come across the terms variate, variable, random variable and be somewhat bemused by what they mean and when to use one and not the other. Such confusion is hardly surprising. Alder and Roessler[1] define a variable as "a measurable characteristic" and a "variate" as an "individual measurement of a variable". Whereas F. H. C. Marriott[2] uses the term variate, or random variable, to describe a variable where the possible values (of the variable) follow a probability distribution. In this text we have used the terms variable or random variable and referred to specific value(s) of a variable as "observed value(s)".

1.4   A Note for Non-mathematicians

Some readers may wish to develop their grasp of subjects mentioned in this account. Such ambition is admirable but not, however, particularly easy to achieve. One begins with an elementary textbook and progresses to more demanding works. The process is rather like climbing a mountain: at first all goes well, one becomes a little breathless but is still going upwards with some confidence. Then the slope steepens and one finds oneself hanging on by one's fingertips. Breathlessness increases but we are undaunted. Sooner or later some ghastly overhang confronts us or there is a lack of foot-holds and we are unable to get any higher. The fingertip stage is characterised by a lack of familiarity with the processes of manipulation: one lacks the facility for juggling with equations and combinations of symbols. This difficulty is surmountable but it takes time. The ghastly overhang–no footholds stage is characterised by a lack of knowledge of more advanced mathematics, especially of something more than elementary calculus. This is, for some of us, the end of our climb towards intellectual satisfaction. One can detect the onset of this stage when, in textbooks, one first meets the phrase "the proof has been omitted". Such blocks to progress are frustrating and one looks for a more advanced textbook. Such books are easily found but "the proof" may appear to be incomprehensible because it involves mathematical techniques of which one has not heard. The feeling of intellectual frustration increases. The amateur climber of mathematical mountains must, for his or her own intellectual comfort, accept that not everybody has the special talent needed to become a mathematician. One's comfort is increased by recalling that one wants to use the results of mathematical endeavour rather than to discover these results oneself and that if one wants to see what's on the other side of the mountain one might go round it rather than over it. This, of course, is no excuse for a lack of knowledge of the results of the labours of mathematicians! In this short book we shall accept these results and try to apply them to the problems we face in the air pollution field.

References

1. H. Alder and E. Roessler, *Introduction to Probability and Statistics*, WH Freeman and Company, San Francisco, 5th edn, 1972.
2. F. H. C. Marriott, *A Dictionary of Statistical Terms*, Longman Scientific & Technical, New York, 5th edn, 1990.

2 Logarithms

2.1 Definition

The logarithm of a chosen number is the power to which another number, the base, must be raised to equal the chosen number. Put another way a logarithm is the "opposite" of a power: So, if

$$a^x = b$$

$$\text{then, } \log_a b = x$$

Or, in words, log to the base a of chosen number b equals x.

The chosen number must be (1) real, (2) positive. The base must be a real number. Both the chosen number and the base may be rational or irrational numbers.

Real numbers are numbers that can be expressed on a scale or by an infinite decimal expansion. Real numbers include integers or whole numbers, and rational and irrational numbers. A rational number is one which can be expressed, precisely, in terms of a fraction, for example 0.75 = ¾ or 7 = 63/9; an irrational number cannot be so expressed. The reader will know that π (pi) is an irrational number: 3.14159265.... The reader will also know that the fraction 22/7, used for π in elementary work, only appears to solve the problem of irrationality: 22/7 = 3.1428571 which is a little larger than π. Real numbers are distinguished from imaginary numbers. $\sqrt{-x}$ is an imaginary number in that squaring a number, be it positive or negative, inevitably produces a positive number *e.g.* $(2)^2 = 4$ and $(-2)^2 = 4$.

Basic Mathematics for Students of Air Pollutants
By Robert Maynard and Richard Atkinson
© Robert Maynard and Richard Atkinson 2024
Published by the Royal Society of Chemistry, www.rsc.org

We shall soon meet another irrational number, e, which is the base of Napierian or natural logarithms. There are two systems of logarithms in general use: common logarithms with base 10 and natural logarithms with base e. Common logarithms are also known as Briggsian logarithms.

As mentioned in Chapter 1, log tables and/or slide rules were once widely used and textbooks of elementary mathematics devoted pages to explaining how such aids to calculation should be used. All has changed and a "scientific" calculator will, instantly, give the common logarithm and natural logarithm of any real and positive number. It is still, however, important to know the principles underlying the use of logarithms and these are summarized below as a series of algebraic statements and proofs.

Logarithm of 1:

$$\text{For any base } a, \log_a 1 = 0$$

$$\text{Since, for any } a, a^0 = 1$$

$$\text{Note also} : \log_a a = 1$$

$$\log_a 0 \text{ is undefined}$$

Logarithm of b^n:

$$\text{Definition} : \log_a b^n = n\log_a b$$

Derivation:
Let

$$\log_a b = x$$

$$b = a^x$$

$$b^n = a^{nx}$$

$$\log_a b^n = nx = n\log_a b$$

Logarithm of a product (multiplication):

$$\text{Definition} : \log_a bc = \log_a b + \log_a c$$

Derivation:
Let

$$\log_a b = x, \text{and} \log_a c = y$$

$$b = a^x \text{ and } c = a^y$$

$$bc = a^x a^y = a^{x+y}$$

$$\log_a bc = \log_a a^{x+y} = (x+y)\log_a a$$

Since

$$\log_a a = 1, \log_a bc = x+y$$

$$\log_a bc = \log_a b + \log_a c$$

Logarithm of fractions (division):

$$\text{Definition}: \log_a \frac{b}{c} = \log_a b - \log_a c$$

Derivation:
  Let

$$\log_a b = x, \text{and} \log_a c = y$$

$$b = a^x, \text{and} c = a^y$$

$$\frac{b}{c} = \frac{a^x}{a^y} = a^{x-y}$$

$$\log_a \frac{b}{c} = x - y = \log_a b - \log_a c$$

Logarithm of $1/b$:

$$\text{Definition}: \log_a \left(\frac{1}{b}\right) = -\log_a b$$

$$\text{Derivation}: \log_a \left(\frac{1}{b}\right) = \log_a 1 - \log_a b$$

$$= 1 - \log_a b$$

$$= -\log_a b$$

Since

$$\log_a 1 = 0$$

Logarithm of $\sqrt[z]{b}$:

$$\text{Definition}: \log_a \sqrt[z]{b} = \frac{1}{z}\log_a b$$

$$\text{Derivation}: \log_a \sqrt[z]{b} = \log_a b^{1/z}$$

$$= \frac{1}{z}\log_a b$$

Since

$$\sqrt[z]{b} = b^{1/z}$$

Logarithm, base a, of a^b:

$$\text{Definition}: \log_a a^b = b$$

$$\text{Derivation}: \log_a a^b = b\log_a a$$

$$\log_a a = 1$$

$$\log_a a^b = b$$

2.2 Common Logarithms

Common logarithms are logarithms to the base 10. The common logarithm of a number, b, is written as $\log_{10} b$ or, more often, $\log b$.

A calculator will give the following:

$$\log 2 = 0.30103$$

$$\log 0.2 = -0.69897$$

The second result may come as a surprise if you were taught, in school, that $\log 0.2 = \bar{1}.30103$

You will recall that the number to the left of the decimal point in $\bar{1}.30103$ was described as the *characteristic* and the number to the right of the point, the *mantissa*. You will also recall that the mantissa is always positive but that the characteristic could be either positive or negative and that when negative it was shown with a short line, a bar, over the number. If you had been asked to find $\log 0.2$ you would have replied "bar1 point three zero one zero three". You would have been correct. But the use of the "bar system" was a dodge which allowed a single log table to be used for numbers greater and smaller than 1. This dodge is not needed by calculators. You will have noticed that $\overline{1}.30103$, that is $(-1 + 0.30103)$, equals -0.69897. The calculator gives the logarithm; the log table gives only the mantissa and leaves deciding the characteristic to you. Of course, this is not difficult when working to the base 10: for numbers >1 and <10 the characteristic is 0, for >10 and <100: 1, >100 and <1000: 2 and so on. Some momentary difficulty might be met with numbers less than 1.0. The difficulty lies in the fact that numbers from 1 to 10 have a characteristic of 0 but numbers from 0.1 to 1.0 have a characteristic of -1 and, of course, a positive mantissa.

In the days of log tables most people used four or five figure logarithms. Larger books of seven, or more, figure log tables were available. Simple calculators will give logarithms to seven figures or more.

2.3   Antilogarithms

Let us mention antilogarithms here. The antilogarithm is just the reverse of the logarithm and is usually abbreviated antilog or alog, we could write $alog_{10}$ but the 10 is usually omitted. If the base were, say, 3 (very unlikely!) we should write $alog_3$. We can find the antilog of a number by entering the number and pressing INV log on a calculator. Or we can look up the number in a table of antilogarithms, or look for the number in the body of a table of logarithms (the reverse of looking up a logarithm).

2.4   Natural Logarithms

The base of natural logarithms is the irrational number e (often referred to as Euler's number). What a very odd number to choose for the base! Of course, there are good reasons for choosing e; let us just say that natural logarithms crop up again and again in higher mathematics, that is in the calculus as a result of the appearance of exponential series in derivatives and integrals. For the interested reader we have provided further information on e, its origin and utility in Appendix 2.1. Other than the fact that the base of natural logarithms is 2.718 (expressed to four significant figures) rather than 10, natural logarithms are much like common logarithms.

The natural logarithm of a number, for example 3, is written as ln 3 = 1.0986. When dealing with common logarithms we wrote antilog for the reverse of log(log 100 = 2, antilog 2 = 100). We could write antiln for natural logarithms, but this is never done. If the natural logarithm of a number is x, then the antilog is e^x sometimes written exp(x). Here the base of the exponentiation (see Chapter 1, Section 1.2.1) is e.

$$\text{For example} : \ln 3 = 1.0986 \text{ then } e^{1.0986} = 3$$

Two further points are worth noting.

First, it is more difficult to determine the characteristic of natural logarithms than of common logarithms: everybody can think in terms of orders of tens but orders of 2.718s are less easy. This difficulty is got around by arguing:

$$\ln 1234 = \ln\left(1.234 \text{x} 10^3\right) = \ln 1.234 + \ln 10^3$$

The table of natural logarithms tells us that ln 1.234 = 0.2103 and that ln 10^3 = 6.9078.

$$\ln \ 1234 = 0.2103 + 6.9078 = 7.1181$$

Tables of natural logarithms are a little different from tables of common logarithms in that the values for 10^1 to 10^{10} and for 10^{-1} to 10^{-10} are given as a footnote to the tables. Of course, this will not worry us: the calculator will give us the natural logarithm of any real and positive number we care to enter.

The second point is that the "bar system" is not used for natural logarithms: ln 0.001234 = −6.6975.

2.5 Conversion from One System of Logarithms to Another

It is occasionally useful to be able to convert common logarithms into natural logarithms and *vice versa*. The following series of equations explains how this can be done.

$$\text{Let} \log_b n = x$$

$$n = b^x$$

$$\log_a n = \log_a b^x = x\log_a b = \log_b n . \log_a b$$

$$\log_b n = \frac{\log_a n}{\log_a b}$$

To make this a little more real let us convert ln 1234 to log 1234, recalling that the base of natural logarithms is 2.718 and the base of common logarithms is 10. Remember log base e (2.718) is written ln and log base 10 is simply log.

$$\log 1234 = \frac{\ln 1234}{\ln 10}$$

$$\ln 12324 = \log 1234 \times \ln 10$$

$$\ln 1234 = 3.0913 \times 2.3025$$

$$\ln 1234 = 7.1177$$

We shall make good use of our knowledge of logarithms in later chapters of this book.

Appendix

A2.1   The Number 'e'

The number "e", often referred to as "Euler's number", is an important constant in mathematics. It was first discovered in the 18th century. Like pi (symbol π), it is an irrational number, that is, a number that cannot be written as a simple fraction and therefore has an infinite number of non-repeating digits after the decimal point. In 1748, Euler calculated e to 18 digits; now, using modern computers and computational techniques it has been calculated to trillions of digits. Modern calculators will store e to many digits but if you wish to remember e to a few decimal places, 2.71828 will suffice.

We have come across e in the context of logarithms. We will come across it again when we consider exponential curves. It pops up in other areas too and it is useful to look at some of these to gain an appreciation of the importance and utility of *e* in mathematics and science.

A2.1.1 Compound Interest

Let's start with a topic close to us all: money.

If you have money to invest in a bank or building society savings account, you will be offered interest on your deposit. Often one can choose whether interest is paid annually or monthly. Let's look at this more closely by considering a hypothetical example.

Let's say you have £1000 to invest in a savings account. A generous bank offers an annual interest rate of 10%.

After 1 year, your savings will have increased to £1100 (let's ignore taxes for now): the initial investment plus £100 interest (10% of £1000).

But, what if we were able to have interest added in two six-month intervals: 5% after the first six months and 5% after the second six months. Let's see how that works out.

After six months we will have £1050, our initial investment plus 5% interest. After the second six-month period we will have £1102.50, 5% interest applied to our balance at the end of the first six-month period. This is a little more than when we chose annual interest.

What happens if we take the interest monthly?

At the end of 12 months our total savings will be £1104.71, almost £5 more than for annual interest.

Let's look more closely at the calculations involved:

Month 1 $\quad 1000 + 1000 \times 0.10/12 = 1000(1 + 0.10/12) = 1008.33$ NB: 10%
$$\equiv 0.10$$

Month 2 $\quad 1008.33 + 1008.33 \times 0.10/12 = 1008.33(1 + 0.10/12)$
$$= 1000(1 + 0.10/12)(1 + 0.10/12) = 1000(1 + 0.10/12)^2 = 1016.74$$

Month 3 $\quad 1016.74 + 1016.74 \times 0.10/12 = 1016.74(1 + 0.10/12)$
$$= 1000(1 + 0.10/12)^2(1 + 0.10/12) = 1000(1 + 0.10/12)^3 = 1025.21$$

And so on....

Month 12 $\quad 1000(1 + 0.10/12)^{12} = 1104.71$

So, we earn more interest if we take our interest monthly rather than annually. What we have done is calculate (discrete) compound interest on the investment. Of course, banks know this and if you request interest monthly, the rate offered is lower than if paid annually!

More generally, if we divide our interest rate r into n equal intervals and apply this rate for each interval with a starting investment of x, we get as our final value:

$$x(1+r/n)^n$$

Without any loss of generality, we can set r to 1. As our time intervals for applying interest get smaller and smaller, *i.e.* n gets bigger and bigger and tends to infinity (∞), the term in brackets tends to 2.71828..., or e. One can check this easily with a calculator – see Table 2.1.

So, as we take ever smaller increments, *i.e.* n gets larger, the *increases* in our investment get ever smaller, converging on a number. This maximum multiplier for our initial investment is e; this is theoretical of course as it assumes infinitely small increments in time (called continuous compounding).

So "e" is the limiting value of the series $(1 + r/n)^n$ as $n \rightarrow \infty$. This is very useful as it enables compound interest to be calculated simply as:

$$F = I\left(e^{i*t}\right)$$

where F and I are the final and initial investment respectively, i the investment rate and t the number of years over the life of the investment. In our example above:

$$F = 1000e^{0.1*1} = 1105.17\left(\text{taking e to be } 2.71828\right)$$

Not exactly the same, but close, the difference arising because one is discrete and the other continuous compounding.

Table 2.1 Compounding factor calculated for "n" time intervals.

n	$(1 + 1/n)^n$
10	0.59374
100	2.70481
1000	2.71692
10 000	2.71815
100 000	2.71827
1 000 000	2.71828
10 000 000	2.71828

A2.1.2   Another (Useful) Place to Find e

Consider the following series of terms:

$$\frac{1}{0!}+\frac{1}{1!}+\frac{1}{2!}+\frac{1}{3!}+\frac{1}{4!}+\frac{1}{5!}+\frac{1}{6!}+\frac{1}{7!}+\ldots = \sum_{n=0}^{\infty}\frac{1}{n!}$$

Note: ! means factorial: $4! = 4 \times 3 \times 2 \times 1$ and $0! = 1$.

Working out each term and adding them up gives us 2.71666. Look familiar? If we added more terms and did some more calculations we would get closer to 2.71828… – e.

You might ask what is so useful about this series. Well, it is a special case of the power series, e^x (with $x = 1$) where:

$$e^x = \sum_{k=0}^{\infty}\frac{x^k}{k!}$$

and $y = e^x$ is the exponential growth (and decay) curve. We shall encounter this curve in Chapter 3.

A2.1.3   Other Applications

As we have seen, e is useful as a base for (natural) logarithms. e is also important in a number of other areas including calculus, probability and trigonometry amongst others. For example, in calculus, e is unique in that the derivative of e^x is e^x; it is equal to its own derivative. In plain words this means the slope of the curve (rate of increase of the function) is equal to value of y at any point. In probability theory it features in the solution to a number of practical situations.

This and other applications of e as well as a brief outline of its history are nicely described in the short article in Nature Physics by Stefanie Reichert.[1]

References

1. S. Reichert, e is everywhere, *Nat. Phys.*, 2019, **15**, 982.

3 Exponential Decay and Semi-logarithmic Graph Paper

3.1 Exponential Relationships

In Chapter 2 we met the irrational number e as the base of natural logarithms. In this chapter we shall consider an equation which includes e and which often crops up in scientific work. All readers will know that a radioactive substance undergoes decay; depending on the substance this may be very rapid or very slow. Some readers will have met the following equation:

$$C_t = C_0 \exp[-kt] \text{ or } C_t = C_0 e^{-kt}.$$

This, of course, means that the amount of the substance, C, at some time, t, denoted C_t, can be determined if the original amount of substance, C_0, and a constant, k, are known. The constant k varies from substance to substance. We could also use the equation to describe the disappearance from the human body of some substance introduced into the blood stream or deposited in the lung. The minus sign in the equation implies that the amount of material present decreases with time. Again, most readers will have heard of the half-life of radioactive materials. This is the time over which the amount of material present declines to half its original value. There is room for a schoolboy error here: on being told that the half-life of a substance was two days he said, "I understand that: half of it disappears in two days and the other half in the next two days." He was only half right. Half of the material

Basic Mathematics for Students of Air Pollutants
By Robert Maynard and Richard Atkinson
© Robert Maynard and Richard Atkinson 2024
Published by the Royal Society of Chemistry, www.rsc.org

does disappear in two days, but a *half of what is left* disappears in the next two days, and a *half of what is then left* disappears in the next two days and so on and on. This progressive halving is the key.

Let us rewrite the equation putting $\dfrac{C_0}{2}$ for the amount of material at time $t_{1/2}$. Note that $t_{1/2}$ means the half-life of the material and NOT half of some yet to be defined period of time, t.

$$\frac{C_0}{2} = C_0 \exp\left[-k \cdot t_{1/2}\right]$$

Divide all through by C_0

$$\frac{1}{2} = \exp\left[-k \cdot t_{\frac{1}{2}}\right] \tag{3.1}$$

Let us now take the natural logarithm of the left and right sides of eqn (3.1). In handling the right-hand side of eqn (3.1) we need to recall that

$$\log_a a^b = b$$

$$\ln\frac{1}{2} = -k \cdot t_{1/2}$$

Dividing all through by $-k$ yields

$$-\frac{1}{k}\ln\frac{1}{2} = t_{1/2}$$

Now then, $\ln\dfrac{1}{2} = -0.693$, and, $\left(-\dfrac{1}{k}\right) \cdot (-0.693) = 0.693/k$

Therefore

$$t_{1/2} = \frac{0.693}{k}$$

If we know the half-life we can, therefore, calculate k.

3.2 Application to Real Data

Let us now look at some data on the disappearance of a substance injected into the blood stream of an animal. Blood samples were taken at specified time intervals and the concentrations of the substance determined: the time points and concentrations are shown in Table 3.1. These data are taken, with permission, from "Mathematics for Biologists"[1] – see also Further Reading.

It is obvious that the concentration is declining with time. It is also obvious that the rate of decline is slowing as the concentration falls.

Let us draw a graph of the data: we shall use ordinary axes, that is axes with arithmetic scales; see Figure 3.1.

It is clear from the data points alone that the relationship is a curve. The general equation for an exponential relationship is:

$$y = A \cdot \exp[mx]$$

where A and m are constants which vary from case to case. Figure 3.1 also shows this curve fitted to the data (we omit details of the statistical method used to estimate these parameters). The estimated values for A and m are 19.6 and −0.053 respectively. Note that m is negative, indicating decreasing values of concentrations as time increases.

Table 3.1 Blood concentration of a substance at specified time intervals. Data from ref. 1

Time (minutes)	0	10	20	30	40	50	60
Concentration (μg cm^{-3})	20.1	10.3	6.5	4.7	2.6	1.8	1.2

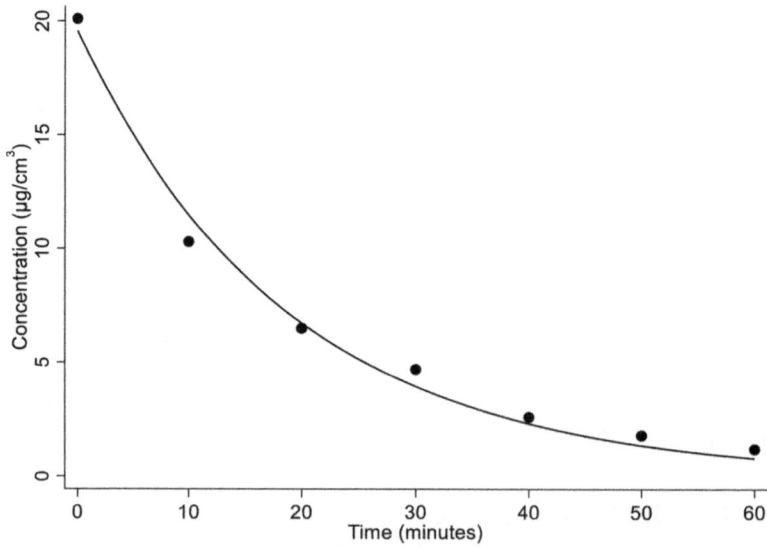

Figure 3.1 Plot of blood concentration of a substance (*y*-axis) at specified time intervals (*x*-axis). Observed values are represented by solid dots. The solid line shows the fitted exponential curve.

The reader can see the data do not form a perfect curve. This is not unexpected with real data and is due to errors in measurement, random variation, *etc.*

Recall the equation we met when considering radioactive decay

$$C_t = C_0 \exp[-kt]$$

Using this specification, $C_0 = A = 19.6$ and $k = |m| = 0.053$ (recall $||$ means the modulus or absolute value of a number).

We can now write a new equation

$$C_t = 19.6 \exp[-0.053t]$$

Recalling that

$$t_{\frac{1}{2}} = \frac{0.693}{k}$$

$$t_{\frac{1}{2}} = \frac{0.693}{0.053} = 13.1 \text{ minutes}$$

From the statistical model fitted to the data, our estimate of the half-life is 13.1 minutes. Let us check, by inspection of the plot, that this looks correct (it is always worth checking your calculations make sense). In Figure 3.2 we have added vertical lines at 13.1, 26.2 and 39.3 minutes, 1, 2 and 3 half-life time periods. By reading along to the y axis from where these lines intersect the fitted curve we can see that the concentrations halve, approximately, for each half-life time period. Of course, with knowledge of the equation parameters, the concentrations can be calculated directly by plugging into the equation the corresponding values of t. It is also worth mentioning that these numbers are derived from the modelled relationship, rather than actual observations (of which we have only 7). As such, they are estimates of the concentrations at the given time points.

Let us now plot the natural logarithms of concentration against time. Again, we shall use ordinary axes (Figure 3.3).

The points lie, more-or-less, along a straight line. What is the equation of the line?

If we take the natural logarithms of the terms and rewrite the equation we obtain

$$\ln y = \ln A + mx$$

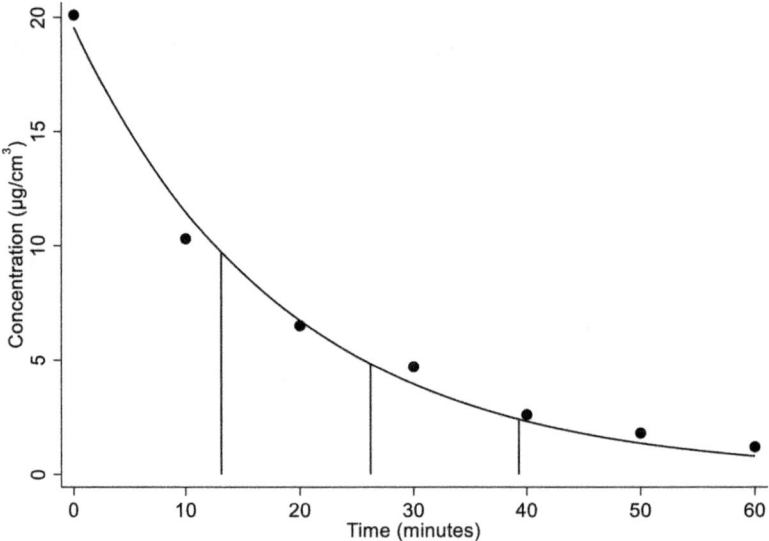

Figure 3.2 Plot of blood concentration of a substance (*y*-axis) at specified time intervals (*x*-axis). Observed values are represented by solid dots. The solid line shows the fitted exponential curve. Vertical solid lines indicate 1, 2 and 3 half-life time periods.

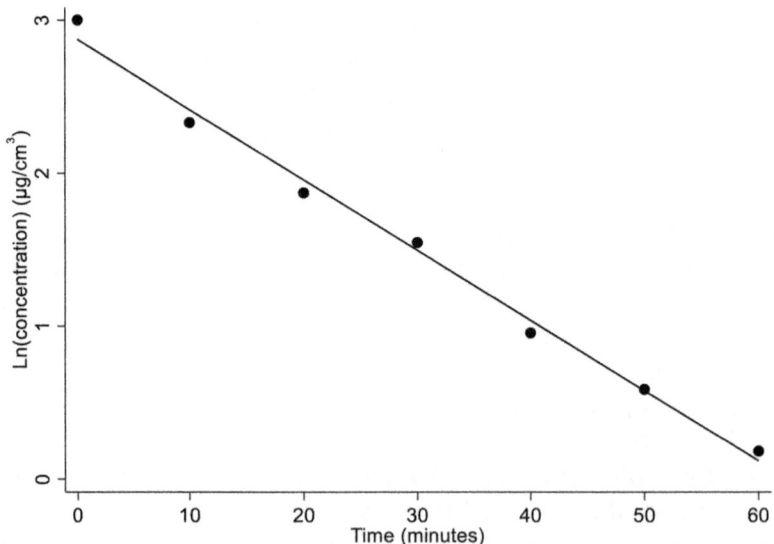

Figure 3.3 Plot of the natural logarithm of blood concentration of a substance (*y*-axis) at specified time intervals (*x*-axis). Observed values are represented by solid dots. The solid line shows the fitted exponential curve.

This is the equation for a straight line. *A* is a constant, so ln *A* is also a constant; *m* represents the gradient or slope of the line. As before, we can estimate the values of *A* and *m* from the data.

We consider line fitting techniques in more detail in Chapter 12.

3.3   Using Semi-logarithmic Graph Paper or Its Equivalent

Let us now replot the graph shown in Figure 3.1 using the original, untransformed, data values using semi-logarithmic graph paper: Figure 3.4. Note the scale on the *x* axis: it is arithmetic, the divisions are equal to one another. The scale on the *y* axis is logarithmic: note that the scale begins, at the lower left corner, with 1 and rises in ten steps to 10, and a further ten steps to 100. Note that the separation of the tick marks decreases as we move from 1 to 10 and from 10 to 100. The advantage of semi-log graph paper is that we do not have to look up the logarithms (natural or common) and can simply plot the data as it stands. Examine Figure 3.4.

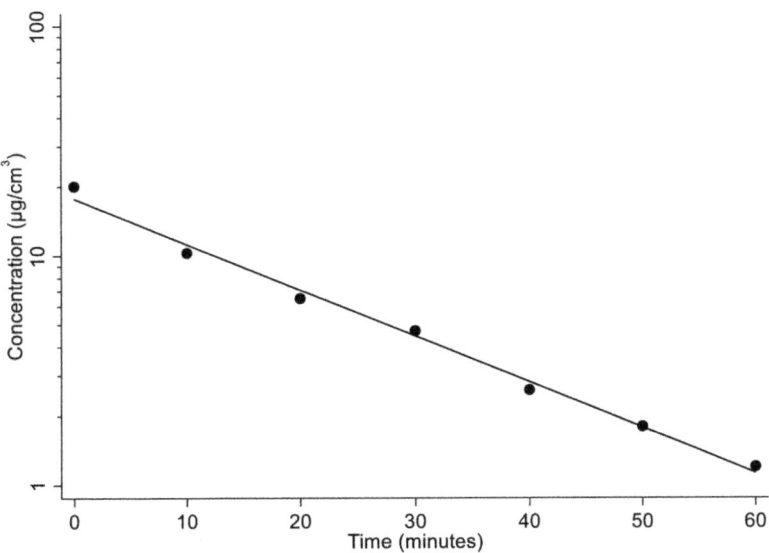

Figure 3.4   Plot of blood concentration of a substance (*y*-axis with logarithmic scale) at specified time intervals (*x*-axis). Observed values are represented by solid dots. The solid line shows the fitted exponential curve.

The importance of understanding natural logarithms should now be clear.

References

1. A. F. Crowe and A. Crowe, *Mathematics for Biologists*, Academic Press Inc., London and New York, 1969.

4 Other Curvilinear Relationships

4.1 Other Exponential Relationships

Consider the equation

$$y = a^x$$

where a is a constant. Here we are concerned with the growth of y as x increases. As x increases: 1,2,3,4... so y will increase: a^1, a^2, a^3, a^4.... This is a multiplicative relationship: each value of y equals the preceding value multiplied by a. By way of example, let $a = 2$. Table 4.1 gives values of y for $x = 1$ to 10.

We can see that y doubles per unit increase in x, irrespective of the initial value of x. The relationship between y and x is multiplicative. Note that absolute changes in y per unit change in x depend on the starting value of x. For example, $(y|x = 3) - (y|x = 2)$ equals 4 ($8 - 4$) whereas $(y|x = 5) - (y|x = 4)$ equals 16 ($32 - 16$). Note, the vertical bar $|$ in the above notation means "given"; for example, $y|x = 3$ means the value of y given that $x = 3$.

From the above data we can see that y starts small but quickly reaches large values. This can cause some difficulty when presenting data in graphical form. Furthermore, it can be advantageous to transform data so that they follow a linear relationship (in statistical modelling for example).

Basic Mathematics for Students of Air Pollutants
By Robert Maynard and Richard Atkinson
© Robert Maynard and Richard Atkinson 2024
Published by the Royal Society of Chemistry, www.rsc.org

Table 4.1 Values of $y = 2x$ for selected values of x.

x	1	2	3	4	5	6	7	8	9	10
$y = 2^x$	2	4	8	16	32	64	128	256	512	1024

Table 4.2 Values of $y = 2x$ and $\ln(y)$ for selected values of x.

x	1	2	3	4	5	6	7	8	9	10
$y = 2^x$	2	4	8	16	32	64	128	256	512	1024
$\ln(y)$	0.00	0.48	0.76	0.96	1.12	1.24	1.35	1.44	1.52	1.60

We may convert the equation above into logarithmic form:

$$\ln y = x \ln a.$$

We may use common or natural logarithms; $\ln a$ is, of course, a constant, $= \ln 2$ in this example.

Let us add an extra row for $\ln y$: Table 4.2.

It will be immediately seen from Figure 4.1 that the plot of $\ln(y)$ against x (hollow circles) is a straight line, whereas the plot of $y = 2^x$ (solid circles) is a curve. For unit increases in x, $\ln(y)$ increases by *a fixed amount* (equal to $\ln(2)$ or 0.69) – this is the slope of the straight line. Whereas y increases by *a factor* of 2 for each unit increase in x. By taking logarithms we have transformed a multiplicative relationship into a linear relationship. The *golden rule* in examining relationships between two variables is DRAW THE GRAPH.

We shall return to this idea of multiplicative and additive relationships in Chapter 10 when we introduce the concept of relative risk. Relative risk is an important measure in the study of air pollution and its effects on health. Understanding how to interpret relative and absolute changes in risk for a given change in air pollution concentration is fundamental in assessing environmental policies aimed at mitigating the health effects of air pollution.

4.2 The Power Equation

Consider the equation

$$y = Ax^n$$

Taking natural logs

$$\ln y = \ln A + n \ln x$$

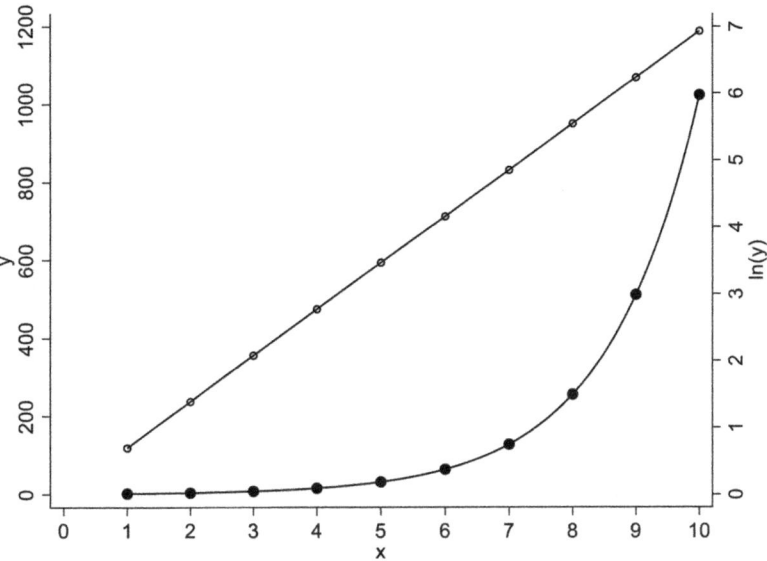

Figure 4.1 Plot of y = 2x (solid circles) and ln(y) (hollow circles) for selected values of x.

or

$$\ln y = n\ln x + \ln A$$

This is in the form of the familiar equation for a straight line

$$y = mx + c$$

where m is the gradient or slope and c the intercept on the y axis.

We could plot a graph of $\ln y$ against $\ln x$ and expect to find a straight line. Rather than looking up the logarithms of x and y we may plot the graph on log–log graph paper. Unlike semi-log paper which we have already mentioned, log–log paper has both axes scaled according to logarithms. A plot of $y = x^2$ on log–log paper is shown in Figure 4.2.

Of course, n in the above equation might be a fraction, for example

$$y = Ax^{\frac{1}{2}} \text{ which means } y = A\sqrt[2]{x}.$$

It is obviously necessary for the log–log paper to be orientated so that $y = 1$ and $x = 1$ appear in the lower left corner of the graph.

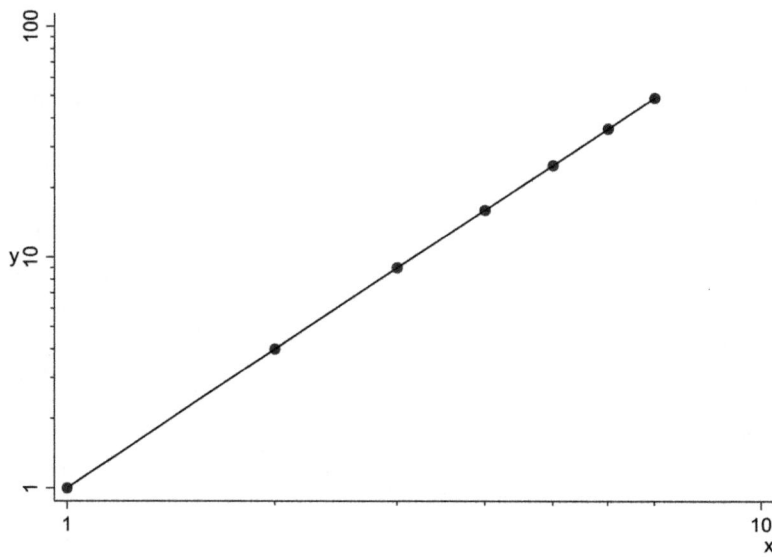

Figure 4.2 Plot of *y* = 2*x* for selected values of *x* using logarithmic scales on both the *y*-axis and *x*-axis.

5 Preliminary Examination of a Set of Measurements

5.1 Introduction

The reader will be familiar with the need to examine a set of measurements; for example, the size of particles collected from an aerosol, the heights of people, the lengths of the legs of insects. Each of these attributes (size/height/length) can be described as *variables* – something that can be measured or counted or recorded for each item/person being assessed. For example, age, height and weight are variables that can be recorded for each person in a survey. In statistical and mathematical notation, a variable is represented by a letter.

The reader will also know that these data (values of a variable) must be arranged so that they can be appreciated and interpreted, that is before information can be obtained from the data. Note: Data is the plural of datum, datum means a single piece of information.

To summarise data, we think about (1) the shape of the distribution of the data points; (2) a measure of central tendency (such as the mean or average value of the data); and (3) a measure of dispersion (such as the variability of the data). These subjects are dealt with in this chapter.

Before we start, it is worth noting that data can be classified as either categorical or continuous. As the name suggests, categorical data are data that can be allocated to groups or categories. Individual items are assigned to one of a number of mutually exclusive categories and the count or frequency, by category, is of interest. Examples are eye

Basic Mathematics for Students of Air Pollutants
By Robert Maynard and Richard Atkinson
© Robert Maynard and Richard Atkinson 2024
Published by the Royal Society of Chemistry, www.rsc.org

colour, sex, particle count by bin size. Continuous data are data that can take any value within a given range: height, temperature and particle mass and concentration for example. The presentation and statistical analyses of data differ for categorical and continuous data. In this chapter we will focus on continuous data.

5.2   Distribution of Data

To summarise the distribution of values we use a bar chart or a histogram. Which we use depends upon the type of data: bar charts for categorical data, histograms for continuous data.

To describe graphically the distribution of a set of measurements on a continuous scale we first divide the range of measurements into a series of intervals or groups (often of equal width) and record the number of measurements that fall into each interval. It is not always possible, or convenient, to make all the intervals of equal width; we shall soon see examples of the problems this may cause. The data can then be plotted as a histogram, as a frequency polygon or as a cumulative frequency polygon; see below for examples. There is no need to labour the point that such plots often suggest that the distribution of measurements conforms to a recognisable curve, often the bell shaped normal distribution curve. Examples of how well many measurements, *e.g.* the heights of people, conform to such a curve are given in all textbooks of elementary statistics. Sometimes the curve that appears on plotting out the data is clearly rather different from a normal curve. Much of statistical analysis is based on the normal distribution curve and methods for transforming the raw data so that their distribution conforms to the normal distribution curve have been developed. There is also a branch of statistics that deals with data that do not conform to such established distributions and provide methods to describe and analyse distribution-free data: non-parametric statistics. We shall discuss summarising such data in the next section.

It is worth remembering that the normal curve, though very widely encountered, is not a fundamental "law of nature". Explanations for the curve depend on the law of errors: the idea, very roughly, being that if some process is subject to random error or variation then the size of individual errors will vary with small errors being more common than large errors. An example is provided by the wear seen on an ancient doorstep: one might expect the feet of those passing through the doorway to be placed in the midline, or thereabouts, but the process of placing is subject to variation and the stone is worn away in a characteristic curve. Yule and Kendall,[1] see Further Reading, point out that the great mathematician Poincaré recorded that a colleague had said to him "everybody believes in the law of errors, the experimenters because they think it is a mathematical theorem, the mathematicians because they think it is an experimental fact". Let us jump ahead a little: it is a fact that even though the pattern of distribution of data in samples taken from a population may be far from normal, the distribution of the means of those samples will conform, closely, to a normal distribution. This very important fact is based on the "central limit theorem"; for discussion see the books on statistics listed in Further Reading.

5.3 The Histogram

It may seem unnecessary to plot a histogram but it is essential for a number of reasons. It shows the range of data, where values are concentrated and whether there are any outliers. It is also important for another reason. The histogram shown in Figure 5.1, shows the distribution of breadths of the heads of 1000 students at Cambridge

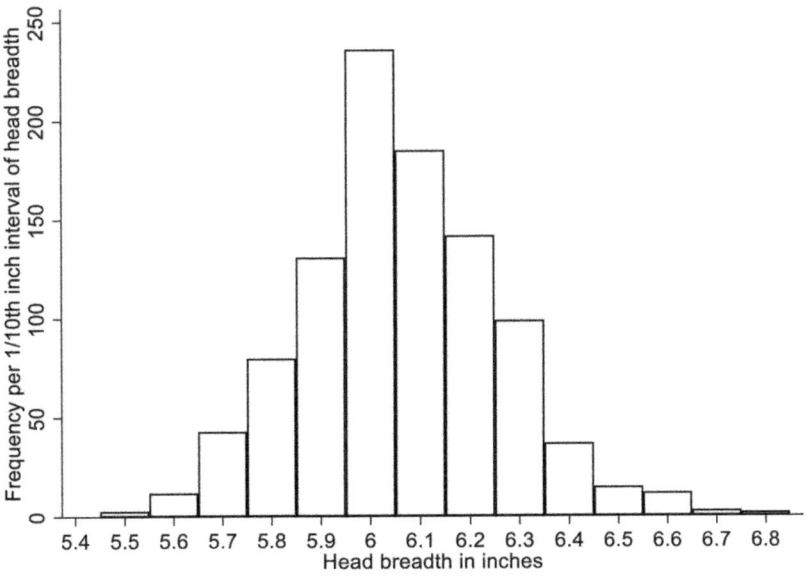

Figure 5.1 Distribution of breadths of the heads of 1000 students at Cambridge University.[1] Data taken from ref. 1.

University in about 1900. The data are taken, with permission, from Yule and Kendall:[1] see Further Reading.

The head breadths, measured to the nearest tenth of an inch, are plotted on the _x_ axis, the abscissa. Look closely at the labelling on the _y_ axis, the ordinate: frequency per tenth of an inch interval of head breadth. In many books the _y_ axis would have been labelled frequency or, perhaps, number. But the precise wording is important: the height of each block is the frequency relating to the associated interval shown on the _x_ axis. We shall return to this point when we consider the distribution of particle diameters in an aerosol.

It does not require much imagination to see that if we joined the midpoints of the tops of the blocks we would produce a frequency polygon – Figure 5.2.

The distribution looks like a normal distribution curve. We can replace the polygon with a normal curve, Figure 5.3.

We might have obtained an even "better" approximation to a normal curve if we had plotted the frequency per tenth of an inch of, say, the heights of a large sample of the male population of the UK. Why? Well, a larger sample from the general population will better represent the full range of values. In Chapter 7, we will introduce a method to assess how well these data conform to the normal distribution.

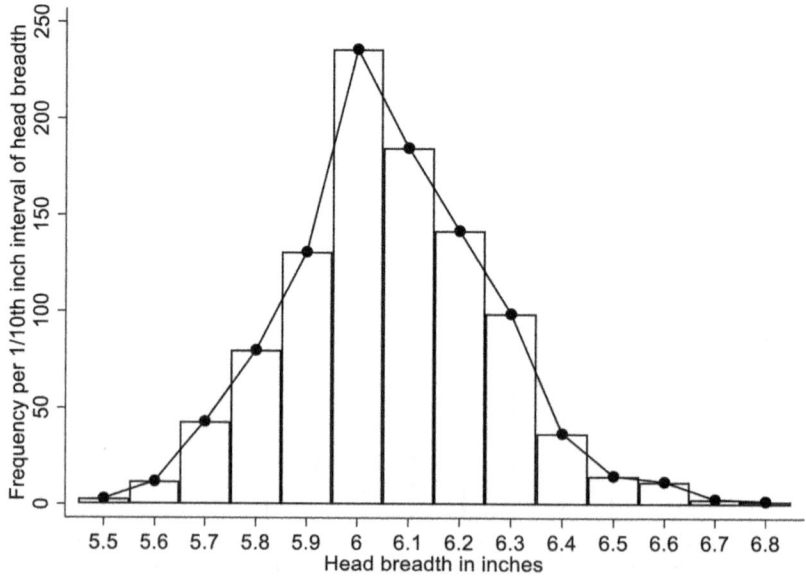

Figure 5.2 Distribution of breadths of the heads of 1000 students at Cambridge University with interval frequencies (solid dots) joined (solid line).[1] Data taken from ref. 1.

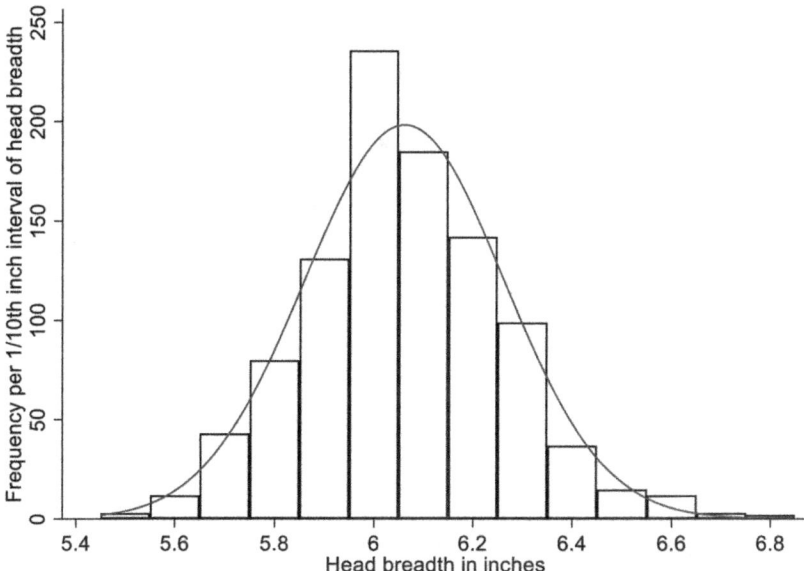

Figure 5.3 Distribution of breadths of the heads of 1000 students at Cambridge University with superimposed normal distribution.[1] Data taken from ref. 1.

However, we have a problem. Given that we have a smooth curve, what does the labelling on the *y* axis actually mean? It was obvious what it meant when we were looking at the blocks of the histogram, but what does it mean when we are looking at the curve? One thing is certain: if we take a point on the *x* axis, say 6.0 inches and draw a vertical line up to the curve, then the frequency we can read off the *y* axis does *not* tell us how many students have heads of exactly 6.0 inches in breath. The thing to remember is that we began with breadths measured to a tenth of an inch and progressed to the blocks of the histogram. The correct answer to the question is: the frequency of a head breadth of *between 5.95 and 6.04* inches (a span of one tenth of an inch) is given by the point on the curve corresponding to a value of 6.0 on the *x* axis. The importance of all this is that we are dealing with frequency per 0.1 inches of head breadth; we must not forget that. We shall return to this point when we consider the normal curve in a little more detail.

We can also plot the cumulative distribution of head breadths: Figure 5.4. The cumulative distribution is simply the running total of *y* values for increasing values of *x*. So, in Figure 5.4, we plot on the *y* axis the total number of people with head breadths up to and including a given value. For the largest head breadth, the cumulative frequency will equal the sample size.

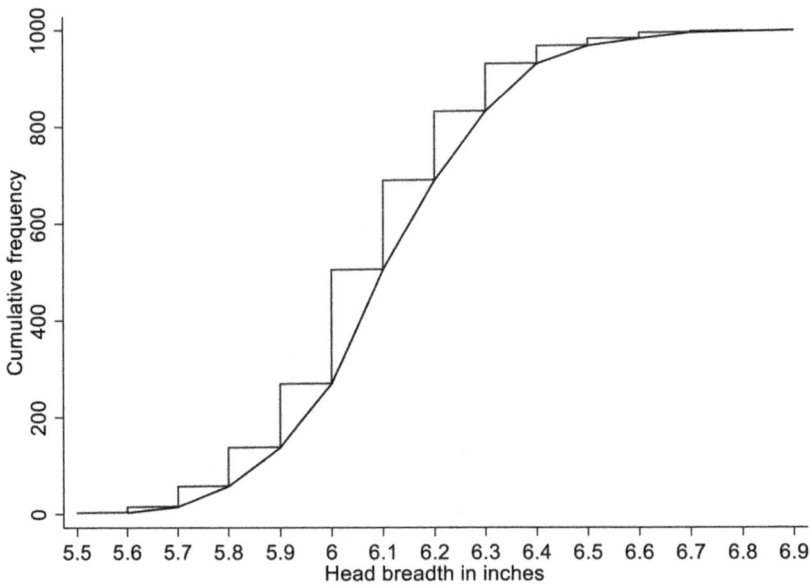

Figure 5.4 Cumulative distribution of breadths of the heads of 1000 students at Cambridge University.[1] Data taken from ref. 1.

Again, the labelling on the *y* axis should be noted. Here, "cumulative frequency" means the total number of people having head breadths of *less* than the specified breadth measured to the nearest 0.1 inches. Such detail is not usually stated: the *y* axis of a cumulative plot is usually labelled "cumulative frequency" or "total number" but the origin of the data, shown in the histogram, should always be recalled. Let us now turn to mathematical methods for describing the distribution.

Returning to Figure 5.1 for a moment, it is worth noting again the scale on the *y* axis – frequency per 1/10th inch interval of head breadth. The axis shows the frequency or number of students in each interval. Hence, the area represented by all of the columns is the total number of students, 1000. A useful, alternative presentation of the data is a plot of relative frequency per 1/10th inch interval of head breadth (*y* axis) against head breadth (*x* axis), the relative frequency being obtained by dividing the number of students in each interval by the total number of students – Figure 5.5. Note the scale of the *y* axis. It shows the proportion of all cases in each interval. In this case the area of all of the columns sum to 1. Such a plot is useful if one is comparing distributions with different numbers of data items, but it also leads us to the relative frequency density plot where the scale on the *y* axis is relative frequency per unit of head breadth. Such a plot

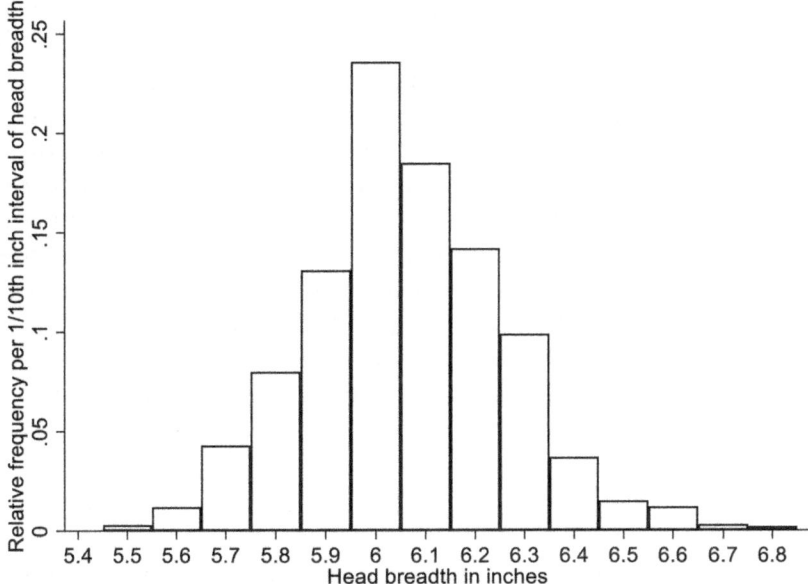

Figure 5.5 Distribution of breadths of the heads of 1000 students at Cambridge University.[1] Data taken from ref. 1.

would be used if the widths were not uniform or were not per unit (*e.g.* per 2 inch head breadth). The relative frequency density is also the probability density function or PDF and we shall come across this in Chapter 6 when we introduce the normal distribution.

5.4 Measures of Central Tendency

5.4.1 The Arithmetic Mean

The most familiar measure of central tendency is the arithmetic mean, or in everyday language "the average". Everybody knows how to calculate an arithmetic mean, or average: add up all the values of the variable (*x*) and divide the total by the number of observations. In the symbols of statistics:

$$m = \frac{\sum_1^N x_i}{N}$$

m, of course, is the arithmetic mean. The right-hand side of the equation tells us to sum (add up) all the *N* values of *x* and divide by *N*. The term x_i implies any value of *x*: the first (*i* = 1), second (*i* = 2), third

($i = 3$)... last ($i = N$) values. The large sigma (Σ) simply means "sum", in this case x, over all values of i with the lower and upper values of i indicated at the bottom and the top of the "sigma": 1 and N. In the old days of working statistical calculations by hand, formulae for calculating the arithmetic mean from grouped data were useful, the computer and calculator have made such formulae obsolete. The arithmetic mean is, in general, the most useful of measures of central tendency: it takes into account all the data, it is easy to calculate, if a series of independent samples are taken from a population, then the arithmetic means of the samples tend to agree rather well with one another. But it also has drawbacks: the most obvious is that because it takes all the data into account, it could be influenced by extreme values of the variable. We return to this in the section on the median.

5.4.2 The Weighted Mean

Consider five independent or separate samples of some variable. Let the numbers of observations in the samples be: 10, 12, 20, 30 and 100. Let the sample means, calculated as arithmetic means of course, be a, b, c, d, e. How should we average these means? We could just average them – add up all five means and divide by 5. This approach implicitly gives equal weight (1/5th) to each sample mean irrespective of the number of observations in each sample. Therefore, the sample of 100 people provides as much information as a sample of 10 people. Intuitively this does not seem the best approach. It makes excellent sense to weight the sample means according to the size of the samples. We can write the weighted mean of the sample means as:

$$\frac{10a + 12b + 20c + 30d + 100e}{10 + 12 + 20 + 30 + 100}$$

Let's look at a simple example to illustrate this. We wish to estimate the mean particle concentration a city's population is exposed to over a given period of time. One part of the city is predominantly residential, let's say 80% of the population lives in this neighbourhood; the other is mainly industrial with the remaining 20% of the population living in this area. Furthermore, the mean particle concentration in the residential area is 4.4 μg m^{-3}, and in the industrial area 28.0 μg m^{-3}.

A simple arithmetic mean is 16.6 μg m^{-3} ((4.4 + 28.0)/2). Is this a fair representation of the mean concentration to which the population is exposed? No, it ignores the fact that most (80%) of the population live in the residential area. The solution is to calculate a weighted mean, giving 80% weight to the mean particle concentration in the residential area and 20% weight to the mean particle concentration in the industrial area. The weighted mean is therefore:

$$0.8 * 4.4 + 0.2 * 28.0 = 9.1 \ \mu g \ m^{-3}.$$

This gives a correct estimate of the mean population exposure to particulates. Note that an asterisk (*) is often used to prevent confusion between the conventional multiplication sign and x meaning the value of a variable.

5.4.3 The Median

The median is the value of x which divides the series, in ascending order, of values of x into two equal parts. For example, consider the following series:

21, 22, 23, **24**, 25, 26, 27

The median is 24.

If there are N values of x and N is an even number, then the median is given by

$$\frac{x_{N/2} + x_{N/2+1}}{N}$$

Consider the following series:

33, 34, 35, **36, 37**, 38, 39, 40

There is now no middle value. The median is taken as the arithmetic mean of 36 and 37: 36.5. What happens if we change the last value to a much larger number: 33, 34, 35, 36, 37, 38, 39, 275?

The median is still 36.5 – it is not affected by the extreme value. However, the mean, now equal to 65.9, is affected.

As the median is unaffected by extreme values or outliers it is clearly not at all dependent on the shape of the distribution of the data. Because of this, statistical tests between medians and other techniques based on ranking the data form a large part of non-parametric or distribution-free statistical methods.

The term non-parametric might appear a little obscure, indeed the word parameter is often, but wrongly, regarded as synonymous with variable. In statistical work the word "variable" refers to a quantity or characteristic that varies from individual to individual (or object to object) and "parameter" to quantity in a theoretical distribution.[2] Based on our inspection of the distribution of the observations in a sample we make assumptions about the parameters of the population distribution. For example, we often assume that the population distribution conforms to a normal curve. Statistical tests based upon such assumptions are described as "parametric"; statistical tests not based on such assumptions are described as "non-parametric". Most textbooks of elementary statistics dwell on parametric methods but non-parametric methods are also important (see books listed in Further Reading for discussion of the two types of method).

5.4.4 The Mode

The mode (modal value of the variable) is that value which occurs most often. In the series just considered there is no mode: each value occurs only once. Now consider the following series

33, 34, 34, 34, 35, 36, 37

The mode is 34. The mode is not of very much value in our work.

5.5 Relations of the Mean, Median and Mode

In a symmetrical theoretical distribution, for example the normal distribution, the mean, median and mode are identical. If the distribution is not symmetrical, that is skewed, these measures of central tendency are not identical. Consider the distribution curves shown in Figure 5.6A and B.

Note the relationship of the mean, mode and median. In moderately skewed distributions these relationships are stable. For example, in a positively skewed distribution the mode is less than the median and the median is less than the arithmetic mean.

5.6 Other Means

The arithmetic mean is familiar but there are two other means which we should consider: the geometric mean and, for enthusiasts, the harmonic mean.

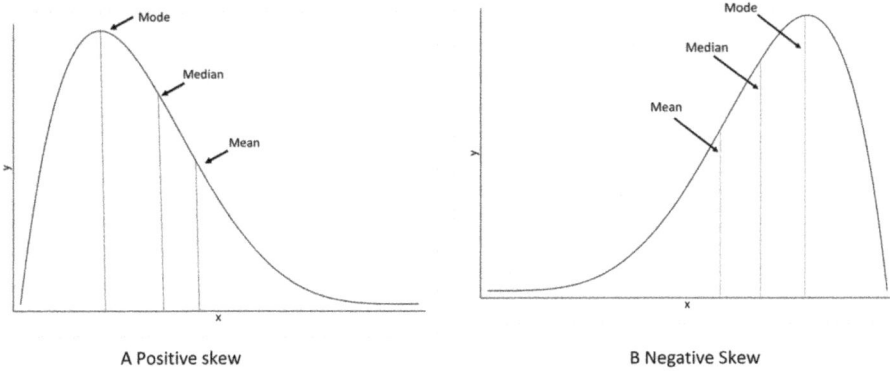

A Positive skew   B Negative Skew

Figure 5.6 Examples of skewed distributions.

5.6.1 Geometric Mean

The geometric mean is calculated as

$$m_g = \sqrt[N]{x_1 \cdot x_2 \cdot x_3 \ldots \ldots x_N}$$

where m_g stands for geometric mean, N is the number of observations, $x_1 \cdot x_2 \cdot x_3 \ldots x_N$ are values of x from first to Nth. Our knowledge of logarithms enables us to write

$$\log m_g = \frac{\log x_1 + \log x_2 + \log x_3 \ldots \log x_N}{N}$$

or

$$\log m_g = \frac{\sum_{i=1}^{N} \log x_i}{N}$$

The geometric mean is used when dealing with data where the ratio between consecutive numbers is constant: a multiplicative relationship, such as we discussed in the previous chapter. We shall return to the geometric mean when we discuss the distribution of particles sizes in an aerosol.

5.6.2 The Harmonic Mean

We include this mean only for fun. Consider the following question: A man drives 100 miles at 20 miles per hour, turns, and drives back at 40 miles per hour. What is his average speed? If, like many people, you would answer 30 miles per hours please consider the mathematician who would argue: 100 miles at 20 mph: 5 hours; 100 miles at 40 mph: 2.5 hours; 200 miles in 7.5 hours = an average speed of 26.7 mph. 26.7 is the harmonic mean.

The harmonic mean is given by

$$\frac{N}{\sum_{i=1}^{N} \frac{1}{x_i}}$$

For the above example: $N = 2$, $x_1 = 20$ and $x_2 = 40$

$$\frac{2}{\left(\frac{1}{20} + \frac{1}{40}\right)} = \frac{2}{\left(\frac{3}{40}\right)} = \frac{80}{3} = 26.7$$

For further information see Further Reading.

5.7   Measures of Dispersion

The most obvious indicator of the dispersion of a set of data is the range: the highest and lowest values of the variable under consideration. The range of the data is always worth examining because it conveys important information. However, as it depends only on the extreme values, it tells us little about the pattern of the distribution. Much more useful are quantiles, which we shall explain shortly.

5.7.1   Variance and Standard Deviation

In this section we shall introduce the variance and standard deviation of a set of data and explain why they are useful indicators of dispersion and how they are calculated. As in all searches it helps to know what we are looking for. We are interested in finding some index of the extent of variation of the data around their mean, some measure, perhaps, of the average extent of variation from the mean. We might take each value of the variable and calculate, by subtraction, how far away from the mean it lies. We might then add up all the variations and average them. Unfortunately, the answer would be zero.

Consider the numbers 3, 6 and 12.

$N = 3$, the sum of the values = 21, the arithmetic mean = 7.

The differences of the values from the mean (mean − value) are 4, 1 and −5: adding gives 0.

We could ignore the signs and add, giving 10, then divide by N, giving 10/3 = 3.33. This is the *mean difference*, that is the arithmetic mean of the differences of the individual values from their arithmetic mean.

Another method would be to square the deviations: 4^2, 1^2, -5^2 giving 42, dividing by 3 to give 14 and then take the square root of 14 to give 3.74. What we have done is to calculate the root-mean-square deviation from the mean of the variable. This example of a root-mean-square deviation has been given a special name: the standard deviation from the mean, or standard deviation (SD) for short. Let us give the formula for SD using conventional symbols.

$$SD = \sqrt{\frac{\sum_{i=1}^{n}(x_i - m)^2}{n-1}}$$

SD is usually shown as σ when we are dealing with a population and as s when we are dealing with a sample (or subset) from a population. The population mean is shown as μ and the sample mean as "m" but we could have written $\bar{x}$ (x bar). The difference between a sample and a population is an important one. A sample is a subset of a population, usually selected at random, such that the sample is representative of the population. Results from analyses of a sample can be used for inference about the population as a whole. For example, consider a general election. When the country goes to the polls, let us assume that everyone who is eligible to vote does so, and the result is the population value. Of course, prior to voting day, numerous polls are conducted with the aim of predicting the outcome of the election. Polls are based on a (small) sample of voters taken from the general population – the result from the sample is then used to predict the outcome of the actual election – this process is called inference. When analysing data, we must therefore be aware, and clear, about what we are calculating – sample means, SDs, *etc.* or population means, SDs, *etc.* For the more mathematically inclined reader we note that when calculating the sample SD, the denominator is $n - 1$, rather than N. "n" is the sample size rather than population size (denoted by N). Also, as we have lost a "degree of freedom" in estimating the sample mean, $\bar{x}$, from the data the denominator is $n - 1$, not n. In practice this is of little consequence, unless "n" is small. The concept of "degrees of freedom" is covered in some detail in Chapter 7 in relation to the Chi-square distribution.

The square of the standard deviation is described as the variance of the data and is written, rather obviously, as σ^2 or as s^2 depending on whether we are dealing with a population or sample:

$$\sigma^2 = \frac{\sum_{i=1}^{N}(x_i - Ł)^2}{N}$$

or for a sample

$$s^2 = \frac{\sum_{i=1}^{n}(x_i - \bar{x})^2}{n-1}$$

The standard deviation is much more widely used than the mean deviation. This is because the standard deviation is easier to handle algebraically and, very importantly, because the normal distribution curve is defined by its mean and standard deviation; see Chapter 6. The normal distribution is symmetrical. When a distribution is skewed, alternative measures of dispersion are appropriate – see Section 5.8.

5.7.2   Standard Error

Much confusion has been caused by the term standard error. It means no more than the standard deviation of the distribution of sample means about the population mean. We can, in fact define two standard deviations:

A: The standard deviation of the values of the variable in a sample around the sample mean – we introduced this in the previous section.
B: The standard deviation of a set of sample means around an estimated population mean.

The second of these, B, is the standard error. Consider a scenario where repeated samples are taken from a population. For each sample we can calculate a sample mean. Let's say we take 100 samples. We have 100 sample means. We can calculate the overall mean of these 100 means; and also their standard deviation. This standard deviation, of the sample means around the mean of the sample means is given a special name, the standard error (SE), to distinguish it from the standard deviation of individual values around a sample mean.

Assuming that we can work only with a single sample and not with the whole population, we can know A but we can only estimate B.

For large samples we assume that $B = \dfrac{A}{\sqrt{n}}$ where n is the size of the sample.

For small samples we assume that B = a modified form of A obtained by substituting $n-1$ for n. We use the formula:

$$SD = \sqrt{\frac{\sum_{i=1}^{n}(x_i - m)^2}{n-1}}$$

and that $B = \dfrac{A_{\text{modifiedform}}}{\sqrt{n-1}}$

The substitution of $n - 1$ for n to produce the modified form of A is well accepted as improving the estimate of B when dealing with small samples: explanations, rather complex, are provided in textbooks of statistics; see Further Reading.

What we must remember

If we are describing a sample, we use A.

If we are describing the population from which our sample was drawn, we use either A, if the sample is large, or the modified form of A if the sample is small.

We use the standard error (B) when calculating the confidence intervals around the sample mean. This, of course, begs the question, what are confidence intervals? We shall return to this question after considering the normal distribution curve in the next chapter.

It is important to note that the SE can be calculated for any statistic derived from a sample – proportion, regression coefficients, *etc.* not just the sample mean.

5.8 Quantiles

The formal definition of quantiles is: The class of $(n - 1)$ partition values of a variable which divide the total frequency of a population or sample into a given number, n, of equal proportions. If we imagine the data arranged in ascending order we could divide the series into quarters (quartiles), tenths (deciles) or into 100 parts (percentiles). Note that in dealing with quantiles we are dealing with numbers of observations not with the values of the observations. Quartiles and percentiles are widely used. One way of representing the spread or dispersion of the data in terms or quartiles is the "box and whisker" plot: this is widely used in the air pollution field: Figure 5.7 shows the box plot of the head breadth data and illustrates a slight skewness.

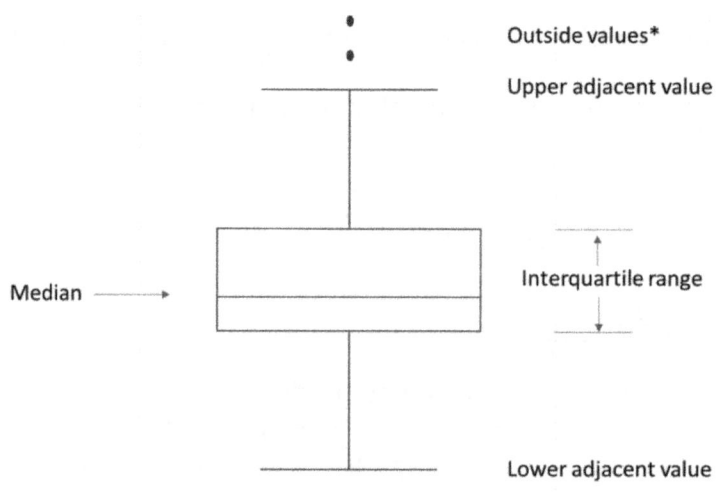

Figure 5.7 Box and whisker plot for breadths of the heads of 1000 students at Cambridge University. Data taken from ref. 1.

Recall that if we have, for example, 100 data points then each quartile will contain 25 data points. The vertical scale represents the data values, the central box shows the location of the central 50% of the data, bounded by the lower and upper quartiles (25th and 75th percentile values), the lines running up and down from the box (the whiskers) show the upper and lower 25% of the data. It is usual to include a bar across the central box indicating the median (50th percentile). Note that the median need NOT bisect the box. The upper and lower bounds of the central box define the inter-quartile range of the data. A moment's thought, as the geometry textbooks used to say, will suggest that if the whiskers are not equal then we are dealing with an asymmetrical or "skewed" distribution. If, on the contrary, we were dealing with a symmetrical, for example normal, distribution the central box would be bisected by the median and the arithmetic mean, and the whiskers would be of approximate equal length. Note some statistical packages present individual extreme values or use other specific criteria to determine the length of the "whiskers".

References

1. G. Yule and M. Kendall, *Introduction to the Theory of Statistics*, Griffin, London, 14th edn, 1958.
2. D. Altman and J. Bland, Statistics notes: Variables and parameters, *Br. Med. J.*, 1999, **318**, 1667.

6 The Normal Distribution Curve

6.1 Introduction

The normal distribution curve is the most important of the several distribution curves met in the study of elementary statistics. As we have already noted the distributions of many variables fit the normal curve rather well. We might lack a clear explanation of why they do so but the fact that they do allows us to use the normal distribution curve to say important things about their distribution.

This is a rather challenging chapter and the reader may wonder why so much emphasis has been placed on the normal distribution curve and on its derivation and characteristics. There are several reasons. First, the distributions of many biological variables follow the normal distribution curve; we have already given an example of this when we considered the distribution of head widths of Cambridge undergraduates in Chapter 5. Second, although the distribution of some variables certainly does not conform to the normal distribution curve, the application of what statisticians describe as "transforms" can convert their distributions into a form that does conform to the normal distribution curve. This is not uncommon in air pollution science where the distributions of particle concentrations and particle sizes are highly skewed. The log transformation is a common choice in such circumstances; we discuss this in greater detail in Chapter 9. Third, a great deal of statistical thinking is based on the normal distribution curve. Last, although the distributions of some variables do not conform to

Basic Mathematics for Students of Air Pollutants
By Robert Maynard and Richard Atkinson
© Robert Maynard and Richard Atkinson 2024
Published by the Royal Society of Chemistry, www.rsc.org

the normal curve the distributions of important statistics that can be derived from the data relating to these variables, such as the means of large samples or the differences between sample means, do conform to the normal curve. This has allowed the development of statistical methods for discovering the likelihood, or probability, of such differences as there may be between sample means occurring by chance.

The reader may now be wondering whether we are limited to the normal distribution. The answer is that we are not. As long as we know the characteristics of a distribution it can be used in essentially the same way as the normal distribution. For example, for small samples Student's "*t*" distribution is used, and for some work the Poisson distribution, an asymmetrical distribution, is used. But the great advantage of the normal distribution is that its characteristics do not depend on the size of the sample, the characteristics of many other distributions do depend on the size of the sample.

6.2   What Is the Mathematical Origin of the Equation of the Normal Curve?

We know, already, that we could sketch in a curve, which looks like a normal curve, given a histogram of, for example, the head breadths of 1000 Cambridge undergraduates (see Chapter 5). But sketching a normal curve does not tell us the equation of the normal curve. That the curve can be described by an equation is certainly true, indeed without this equation we would not be able to determine areas under the curve defined by lines vertical to the x axis, that is, by ordinates of the distribution. We would not be able to calculate the area under the curve defined, for example, by a line drawn vertically from the arithmetic mean of the values of x and one drawn vertically at $x + 1$ standard deviation from the mean.

If the reader has read the note to non-mathematicians given in Chapter 1, they will guess that we are approaching the "clinging by fingertips" stage of mathematical mountain climbing and will soon be facing an overhang which we shall/may not be able to surmount. Let us take a look at that overhang. Those who want to turn back and go round the mountain should do so now.

Yule and Kendall[1] approach the normal curve as a limiting form of the binomial distribution for a large number of observations. Lost already? Let's regroup. The binomial distribution can be determined from first principles. Central to the binomial distribution is the concept of a binary outcome to a "trial": success or failure. Success may be obtaining a head on a toss of a coin, getting a six on a throw of a die, drawing an ace from a pack of cards. Let's proceed using the example of tossing a coin. Assuming the coin is fair – the chance, or probability, of getting a head is 0.5, a tail, also 0.5. The sum of the probabilities is, of course, 1. What if we tossed the coin twice? What possible outcomes are there? The answer is four: a tail and then another tail, a head and then a tail, a tail and then a head, and two heads. The probabilities of getting two tails, one head and one tail (assuming we are not interested in the order in which we get them) and two heads are ¼, ½, and ¼.

The sentence above is an important one. It starts "The probabilities...". In making this statement we have assumed that the process of tossing the coin twice has been repeated multiple, let's say N, times allowing us to work out the frequency with which the event of interest occurs in the long run. For example, if $N = 1$, then the outcome of tossing the coin twice might be two heads, repeat the process and we might get one head and one tail, *etc.* Each time we realise one outcome. If we repeated the process again and again, we would obtain multiple instances of each outcome and the proportion of times (or probability) we would get two heads would tend to ¼; for one head and one tail, ½ and two tails, ¼.

Now then, tedium aside, we could make the calculations for a larger number (n) of trials repeated N times and draw a histogram. As we added to the number of trials, we would find that the histogram would remind us more and more forcibly of the normal distribution curve. This is illustrated in Figure 6.1 which shows the probabilities of obtaining k successes among n trials repeated N times for $n = 1, 2, 5,$ 10, 20, 50 with a probability of a success for each trial of $p = 0.7$ (such as a biased coin, for example). Note, the scale on the y axis and, also, that the sum of all of the bars in each plot is 1.

Note that as n increases the curve become more and more like the normal distribution curve. We'll now consider the normal distribution curve in more detail.

Let us begin by considering a theorem stated by Alder and Roessler (see Further Reading).

In the binomial distribution for N samples (observations) of n trials each where the probability of success in a single trial is p, probability of failure $1 - p$, often denoted q, if the value of n is increased, the histogram approaches a curve, called the normal curve, whose equation is

$$Y = \frac{N}{\sigma\sqrt{2\pi}}\,e^{-(X-m)^2/2\sigma^2}$$

(6.1)

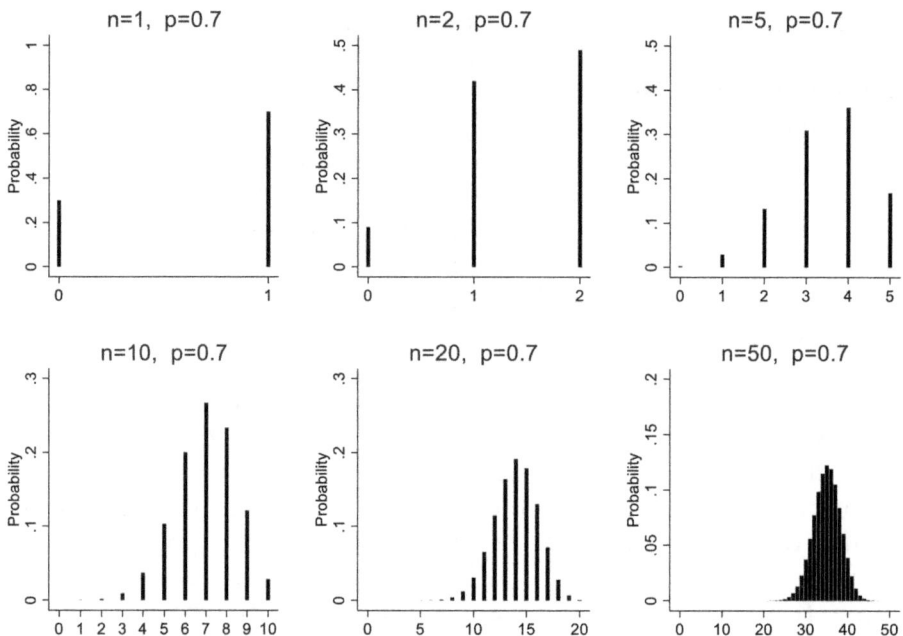

Figure 6.1 Distribution of probabilities of obtaining *k* successes among *n* trials repeated *N* times for *n* = 1, 2, 5, 10, 20, 50 with a probability of a success for each trial of *p* = 0.7.

where *X* = outcome of *n* trials; *m* = mean of the binomial distribution = *np*; σ = standard deviation of the binomial distribution = $\sqrt{npq}$; e = base of natural logarithms = 2.71828...; π = 3.14159...; *Y* = frequency with which any *X* occurs in *N* repetitions of the *n* trials.

If we return to our simple example of throwing a coin twice and counting how many heads we get then *n* = 2, *p* = 0.5, *q* = 0.5, *m* = 2 × 0.5 = 1, $\sigma = \sqrt{2 \times 0.5 \times 0.5}$ and *N* is the number of times we repeat the process of tossing the two coins. So, the formula tells us how many times we will get a particular outcome *X* (*e.g.* 1 head from tossing a coin twice *N* times).

Alder and Roessler point out that they omitted the proof of the theorem; the reader who is interested in following the proof is referred to Yule and Kendall (see Further Reading). The proof involves a good deal of algebra and what may be unfamiliar theorems, but it begins with the binomial distribution and ends up with the equation for the normal curve.

If *Y* is the frequency with which any *X* occurs then it follows that the *relative frequency* (that is the probability, *P*) with which any *X* occurs is given by

$$P = \frac{Y}{N}$$

If we now divide both sides of the equation for the normal curve by N we obtain

$$P = \frac{1}{\sigma\sqrt{2\pi}}\, e^{-(X-m)^2/2\sigma^2} \tag{6.2}$$

This equation is defined as the equation of the *normal probability curve*. It is also called the probability density function (PDF) of the normal distribution. All we have done is to convert from frequency to probability. If we add up the probabilities of all the possible values of X the answer will be 1.0. If we think back to a histogram, we can see that what we are saying is that the total area under the normal curve expressed by the equation is equal to 1.0. With a little imagination we can also see that we might be able to calculate the area under the curve between specified values of X. Unfortunately, an inconvenient feature of the normal PDF (and others for that matter) is that it is not possible to integrate it in order to find areas under the curve. However, areas can be found numerically. To do so for every possible value of m and σ is far from practical. They have though been calculated, tabulated and published for the normal distribution with mean 0 and standard deviation 1 – see Table A1 in the Appendix. To use them we simply need to rescale X to Z, by subtracting m and dividing by σ. Let's see how the rescaling of the equation develops.

Let's make a further modification to eqn (6.2): multiply both sides by σ

$$P\sigma = \frac{1}{\sqrt{2\pi}}\, e^{-(X-m)^2/2\sigma^2} \tag{6.3}$$

What does $P\sigma$ mean? It means that for a plot of the normal distribution the vertical y axis has been rescaled from P to $P\sigma$, so that σ and not 1 is the unit of measurement.

Let us call the composite term $P\sigma$, "y" and write our equation for the normal curve as

$$y = \frac{1}{\sqrt{2\pi}}\, e^{-(X-m)^2/2\sigma^2} \tag{6.4}$$

Now let us look at the right hand side of the equation, let us consider the term

$$(X-m)^2 / 2\sigma^2$$

Let us write

$$z = \frac{X-m}{\sigma} \qquad (6.5)$$

What we have done is to express the difference between a given value of the variable, X, and the mean, m, in terms of sigma which, of course, is the standard deviation. z is referred to as the "normal deviate."

Then

$$z^2 = \frac{(X-m)^2}{\sigma^2}$$

Note that when we squared the left hand side of the equation we also squared the numerator and denominator on the right hand side.

And, dividing all through by 2

$$\frac{z^2}{2} = \frac{(X-m)^2}{2\sigma^2}$$

So, we can substitute in eqn (6.3) for the normal curve and obtain

$$y = \frac{1}{\sqrt{2\pi}} e^{-\frac{z^2}{2}} \qquad (6.6)$$

This equation represents the *standard normal probability curve*.

Given the equation we could calculate y for any value of z but, as indicated, this has been done for us. Values of the ordinates of the normal curve (values of y) for values of z ranging from 0.00 to 3.99 are shown in Table A2 of the Appendix. We also note that from (6.2) and (6.4) the maximum values of P and y are reached when $X = m$ and $z = 0$.

Areas between the midline of the normal curve ($z = 0$) and specified values of z ranging from 0.00 (area = 0.0000) to 3.89 (area = 0.5000) are shown in Table A1 of the Appendix.

If the probability values, for specified values of z are required then we must remember that Table A2 of the Appendix gives values of y for specified values of z and that $y = P\sigma$. So, to obtain P we must divide the value of y provided by the table by σ.

We shall find that being able to determine the area under the curve between two specified ordinates is more useful that being able to determine the ordinate of the curve for a specified value of x. Let us look in a little more detail at how to determine areas under the curve by using tables: see Figure 6.2A.

The area between the ordinate for the mean, m, and that at the specified value of $x1$ has been shaded, let this area be A. A is the fraction of the total area under the curve. Given that there are N observations, the area A represents NA observations. Half the area of the curve lies to the left of the mean, half to the right of the mean. The fraction lying to the left of the mean is 0.5, the total number of observations is N, therefore $0.5 \times N$ observations lie to the left and right of the mean.

Now consider Figure 6.2B. The fraction of the total area under the curve that lies between the ordinates at $x1$ and $x2$ is the fraction between the mean and $x2$ minus the fraction between the mean and $x1$. Multiplying by N gives the number of observations falling into this interval.

As you may have guessed, computer software will calculate, in a flash, areas under the normal curve for any value of m and σ and for any value of X.

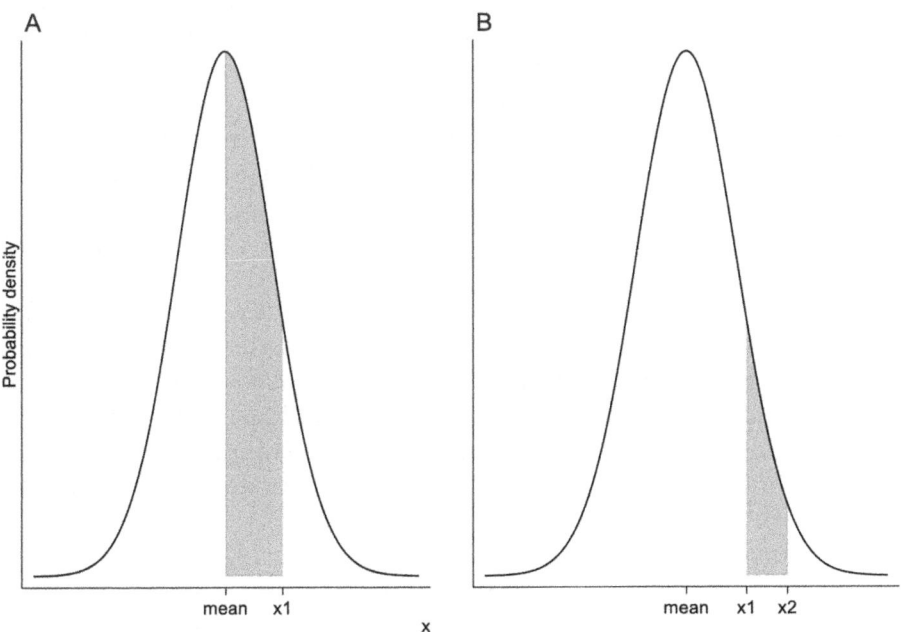

Figure 6.2 Normal probability density curves with the area between the mean and value $x1$ shaded (A) and the area between values $x1$ and $x2$ shaded (B).

Let us conclude this section with two examples of how to use the normal distribution function to estimate the probability of an event and the number, or frequency, of events in a specified interval.

At the start of this chapter we used the binomial distribution as our starting point in the derivation of the normal distribution. From Figure 6.1 we learnt that for a relatively small number of trials, the distribution of binomial probabilities seems to follow quite well a normal distribution. Let's choose $n = 10$ for our example.

From Figure 6.1 we can see that the probability of 7 heads from 10 throws is approximately 2.6–2.7. Using the binomial distribution, the probability is calculated to be 0.267. To estimate this probability from the normal distribution we need two further pieces of information, the mean and standard deviation of the binomial distribution.

In this example, $x = 7$, $n = 10$, $p = 0.7$ and $q = 1 - p = 0.3$ and hence the mean $= np = 7$ and SD $= \sqrt{npq} = 0.1449$.

We can now calculate our standardised z value by substitution of the values for x, m and σ into eqn (6.5). This gives us $z = 0$.

If we find $z = 0$ in Table A1 in the Appendix, the corresponding density is 0.3989. Not what we were expecting from our estimate from Figure 6.1.

But remember, this is the probability density for the standardised normal distribution. To get the density for "our" distribution (which has a mean = 7 and SD = 1.449) we have to rescale the standardised normal density by dividing by the SD from "our" distribution. Revisit eqn (6.3) and hopefully this will make sense.

Hence, the probability of obtaining 7 heads from 10 throws is 0.3989/1.449 = 0.275. Not a bad approximation to the value determined directly from the binomial distribution (0.267).

A calculator or statistical software will of course calculate this value easily using standard functions.

Let us now look at using the normal distribution to estimate the *frequency* of an event in a particular category. That is using the normal distribution as a limit of a frequency distribution of a continuous variable. The normal distribution does in fact provide a good approximation not only for a histogram obtained from the binomial distribution but for many histograms. Note in this scenario we dispense with n and use N to indicate the number of observations.

We return to the head breadths of our Cambridge students introduced in Chapter 5. Figure 5.1 showed the frequency of head breadths of $N = 1000$ students by 1/10th inch interval of head breadth. We superimposed a normal distribution in Figure 5.2. Let's use that distribution to estimate the number of students with head breadth of say 5.9 inches.

How should we approach this? The first point to make is that the column labelled 5.9 is in fact for students with head breadths between (approximately) 5.850 and 5.949. The 5.9 is the centre of an interval. Hence, we are interested not in the probability associated with a single value of head breadth but an interval of head breadth.

Fortunately, we have tables (Table A1 in the Appendix) that tabulate areas under the standard normal distribution. To calculate the area of the column we need to look up, in the table of "areas under the normal probability curve" (Table A1 in the Appendix) the areas between $z = 0$ and $z = z_1$ and $z = 0$ and $z = z_2$, the standardised values of the limits of the column and subtract one from the other. First, however, we need the mean and SD of the 1000 measurements; these are determined in the usual way from the measurements obtained and are 6.0615 and 0.2012 respectively.

Next, we standardised the two measurements that define the ends of the column of interest.

$z_1 = (5.850 - 6.0615)/0.2012 = -1.051$ and $z_2 = (5.949 - 6.0615)/0.2012 = -0.5592$. Do not worry about the minus signs, we can ignore them as the normal distribution is symmetrical about its mean.

We can now look up the areas using Table A1 in the Appendix. Find $z_1 = 1.05$ (the numbers in the table are given to 2 decimal places) in the table, the corresponding area is 0.3531, and for $z_2 = 0.56$, the corresponding area is 0.2123. The difference, by subtraction, is 0.1408. This is the area under the standardised normal distribution between our two z values.

To obtain the corresponding frequency, the number of students with head breadths between 5.85 and 5.949, we multiply by N, the number of observations (1000 students). Hence, the number of students with head breadths in the column labelled 5.9 inches (representing a width of 1/10th inch centred around 5.9 inches) is estimated from the normal distribution to be 141. A quick check against Figure 5.2 shows a discrepancy of 141 *vs.* 131 for this interval.

Again, using a calculator or computer software allows you to calculate this value simply and quickly, which is fine providing you understand what the software is doing for you.

The reader may be wondering why in the first example we had to scale the z value by the SD to arrive at the correct probability density whereas in the second example we did not need to scale the area by the SD. Well, the answer is that in the two examples we are seeking to calculate two different things, one the probability density or ordinate for a specific value of x and in the other an area under the normal distribution curve giving the proportion of observations within a given range of x values. The following figure may help explain this. Figure 6.3

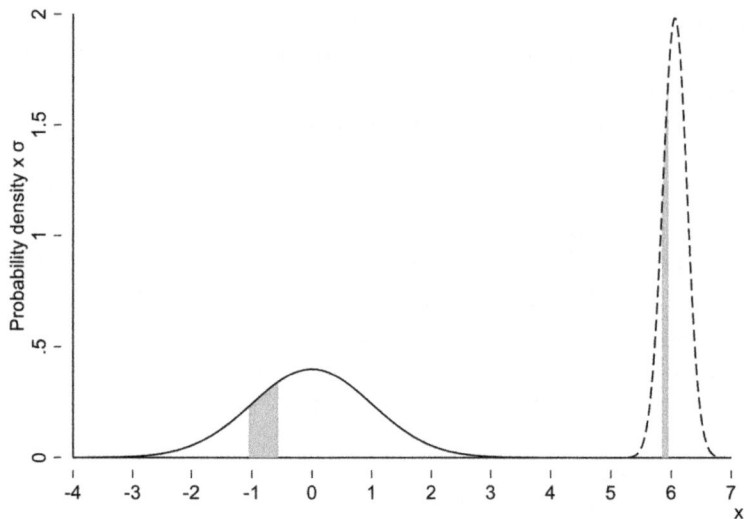

Figure 6.3 Examples of normal distribution curves with different means and standard deviations.

shows the two normal distributions, the original (to the right of the plot) and the standard normal distribution curve (to the left of the plot). The original distribution is centred around the mean of 6.06. The standard normal distribution is, of course, centred around a mean of 0. The SD of the original distribution is 0.2012, much smaller than the standard normal distribution where SD = 1. Hence, the distribution is much narrower, but also taller than the standard normal distribution.

Note, the scale on the y axis in Figure 6.2: 0 to 2. But probability can only take values between 0 and 1. This is true. The reason the curve goes to almost 2 at the mean of the original distribution is because the scale of the y axis in Figure 6.2 is probability density $\times \sigma$ – refer to eqn (6.1).

The SD of the standard normal distribution is 1. So, if we are interested in density for a given x, *i.e.* the height of the ordinate from that x value, then we have to scale the z value to get back to the value of the ordinate for x. If we are interested in the area under the curve and the number of observations between two values of x then it is not necessary to scale by the SD since the transformation from x to z maintains the proportionality of the areas under the two curves – the shaded areas in Figure 6.3 are the same.

One final point for completeness, if two distributions have the same standard deviations but different means then the shapes of the normal curve are the same in both, but they are centred around two different points on the *x* axis (at their respective means).

6.3 Relationship Between Areas Under the Normal Distribution Curve and Standard Deviation

This is most important.

68.27% of the area under the curve is bounded by +1 and −1 standard deviations from the mean.

95.45% of the area under the curve is bounded by +2 and −2 standard deviations from the mean (a common choice (particularly in health studies) is 95%, bounded by ±1.96 standard deviations from the mean).

99.73% of the area under the curve is bounded by +3 and −3 standard deviations from the mean.

We shall make use of these figures when we consider the question of the confidence interval around a sample mean.

6.4 Recap on Progress

Having considered a good deal of theory let us review what we might do in an imaginary piece of work.

1. We collect the data: we are measuring something (for example head breadth or height or blood pressure) in a large sample of subjects.
2. We arrange the data into groups (defined by sensible intervals on the measurement scale) and plot a histogram. Note that if the data are skewed it would be inappropriate to calculate the mean and that a transform (a way of converting a skewed distribution into a normal distribution) might be needed.
3. We calculate the arithmetic mean and standard deviation of the data.
4. Having examined the histogram we think that the data might conform to a normal distribution curve.
5. We know *n*, the number of subjects in the sample, we know the arithmetic mean and the standard deviation. We have all we need to write the equation for a normal curve using not *N* and *z* but actual numbers.

6. We can use tables to find the ordinates corresponding to various values of the measured variable.
7. We can plot the curve.
8. Even without plotting the curve or we can determine areas under the curve and calculate the number of subjects whose measurements lie between A and B, by using values of z, which we will know because we know the arithmetic mean and standard deviation, and reading off the areas.

We might, of course, wonder whether the curve we have adopted actually fits the data. All we have done so far is to look at the histogram, decide that the data seem to conform to a normal distribution curve, and make calculations on that basis. We might wonder how reliable predictions made on this basis actually are. Fortunately, there is a statistical technique that allows us to test the fit of the normal curve to our data. We shall consider this in the following chapter.

We have mentioned tables for the ordinates and "areas under the normal curve" more than once: see tables in the Appendix.

6.5 The Cumulative Plot of the Normal Distribution

There is nothing mysterious about this. We simply plot on the vertical axis the cumulative sum of the probability density, as we progress across the x axis.

The cumulative plot of a normal distribution is a curve, known as an ogive or S shaped curve. Figure 6.4 provides an example.

Note the labelling on the y axis: the probability density, not frequency or number per interval or unit of measurement of x. Of course, we could have plotted the cumulative fraction (or probability) on the y axis expressed as a percentage: the scale running from zero to 100, instead of zero to 1.

The cumulative curve shown in Figure 6.4 has been plotted in the conventional way. In work with aerosols, see later, the axes are reversed, as in Figure 6.5.

6.6 Use of Probability Graph Paper

In this account we shall meet two types of probability graph paper. In both, the y axis is scaled according to the normal distribution curve. In one type, the x axis is a simple arithmetic scale (arithmetic

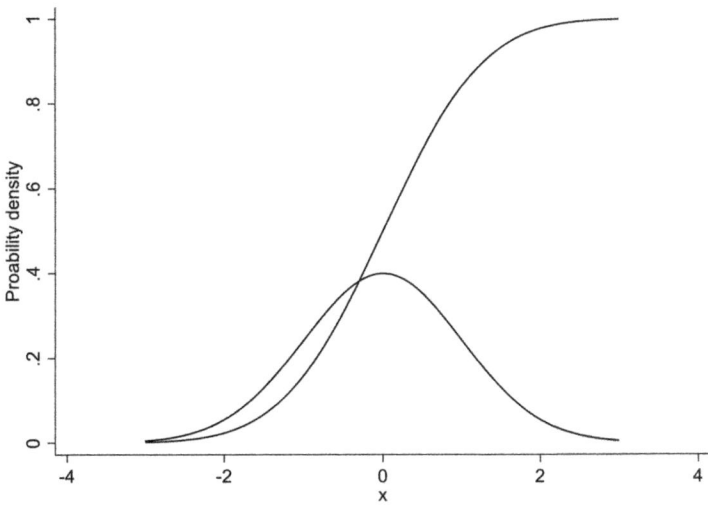

Figure 6.4 Cumulative plot of a normal distribution curve or ogive and normal distribution curve.

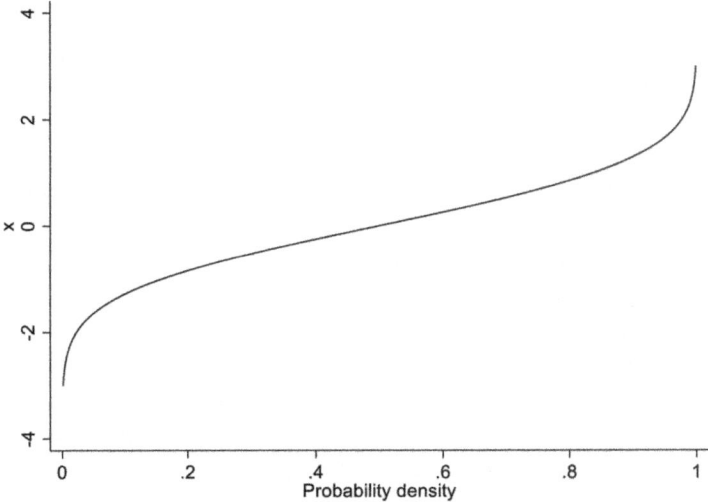

Figure 6.5 Cumulative plot of a normal distribution curve or ogive with axes reversed (from Figure 6.4).

probability paper), in the other the x axis is a logarithmic scale (log probability paper). We shall ignore log probability paper for the moment, although this is, for work on distributions of particle size in aerosols, the more important type.

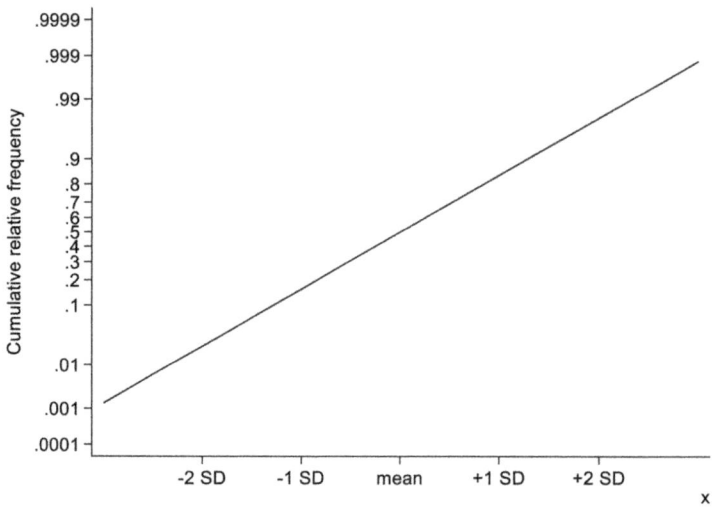

Figure 6.6 Cumulative plot of a normal distribution curve using a probability scale for the *y*-axis.

If the data which when plotted on ordinary graph paper produced an ogive are plotted on arithmetic probability paper a straight line is produced: see Figure 6.6.

It is obvious that we can read off the graph the fraction, or percentage, of the total area under the distribution curve between two points on the *x* axis by simply taking the points on the *x* axis and finding, by examining the line, the related points on the *y* axis. This is a great advantage: we could not do this from what we might call the ordinary normal distribution curve of frequency (or relative frequency as a fraction of 1 or percentages), although, of course, we could define and determine the required area under the curve.

Note that the *y* axis does not run from zero to 1, percent, but from 0.0001 to 0.9999: we know that the tails of the normal curve are asymptotic to the axes and, thus, never actually meet the axes.

We can recap our sequence of exploring the data: data, grouped data, histogram, normal curve, cumulative normal curve (ogive), arithmetic probability plot of cumulative normal curve. Given that our data are normally distributed we are now well placed to use those data. We should ask, does the normal distribution curve fit the data: this is the subject of the next chapter. But before we turn to that, we shall consider calculation of confidence limits of the sample mean based on the normal distribution of sample means around the population mean.

6.7 Confidence Limits of the Sample Mean

We have already said that the distribution of sample means around a population mean is likely to conform to the normal distribution curve. Remember: all we are likely to have to hand, so to speak, is the sample mean and the sample standard deviation. We do not know the population mean, but we know that the means of samples drawn from the population will vary with a standard deviation which we defined, in Chapter 5 as the standard error of the sample mean. Let us calculate the likely extent of this variation. We shall consider an example provided by Alder and Roessler; see Further Reading.

The sample: the number of observations = 100, the sample mean = 40, standard deviation = 11.

The standard error of the sample mean is given by the standard deviation of the sample (11) divided by the square root of the number of observations (100)

$$\text{Standard error of the sample mean} = \frac{11}{10} = 1.10$$

Note, the SE is much smaller than the SD.

We know from our discussion of the normal distribution that 95% of the means of samples taken from the population will lie between about ±1.96 standard deviations (standard errors in this case) from the mean. We need to refine our statement and to do this we use the table of areas under the normal distribution curve. It will be recalled that the table related the measure described as z to the area under the curve. It will be recalled that z is a measure of the distance of some point on the x axis from the mean, expressed in units of standard deviations.

Let us find the value of z for an area of 0.475. Why 0.475? Because 0.475 is one half of 0.95, that is, of the fraction, of the area under the curve, which includes 95% of the means of possible samples. When we look up the value of z corresponding to an area of 0.475 we find that $z = 1.96$. We recall the formula for z

$$z = \frac{x - m}{\sigma}$$

Adapting this to our current argument the limits (or ordinates) of the area which includes 95% of possible sample means is thus given by

$$\frac{\text{limits of the sample mean} - \text{sample mean}}{\text{standard error of the sample mean}} = \pm z = \pm 1.96$$

We can now write

$$\frac{\text{limits of the sample mean} - 40}{1.10} = \pm 1.96$$

or

$$\text{limits of the sample mean} = 40 \pm 2.16$$

The limits of the sample mean are therefore 42.16 and 37.84.

Now we must be careful. It would be an error to say that there is a 95% chance that the population mean lies between 42.16 and 37.84. Why would this be an error? The point is that the 95% probability relates to the probability of the confidence interval containing the population mean, and not to the probability of the population mean lying within the calculated confidence interval. The latter statement is meaningless because the population mean, although unknown, is a finite number, and whether it lies within the confidence interval is either true or false and is not a matter of probability. The confidence interval, on the other hand, is one of a large number of possible intervals because there are a large number of possible samples. Thus, whether the calculated interval contains the population mean is a matter of chance or probability.

We have calculated the 95% confidence interval of a sample mean; we could have calculated the 99% confidence interval by simply looking up the z value corresponding to an area under the curve of 0.495 (half of 0.99). Note also, the process applies to other parameters – it is not confined to means. For example, confidence intervals can be calculated for estimates of a proportion or a regression coefficient. The CI is a measure of the precision of an estimate and applies to any parameter estimated from a sample of data.

The method outlined above applies to large samples: "large" often being taken to mean more than 30. For small samples, a similar approach is taken but the normal distribution curve is replaced with Student's t distribution. Unlike the normal distribution, Student's t distribution changes shape (although not a great deal) as the size of the sample changes; it tends to a normal distribution as the sample size increases. Explanations of use of this distribution are given in textbooks of elementary statistics; see Further Reading. Some readers will know that "Student" was the name under which W. S. Gossett (1876–1937), Head Brewer and statistician to the Guinness Brewery in Dublin, published his scientific work.

References

1. G. Yule and M. Kendall, *Introduction to the Theory of Statistics*, Griffin, London, 14th edn, 1958.

7 Does the Normal Distribution Curve Fit the Data?

7.1 Introduction

We know, from the previous chapter, how to fit a normal distribution curve to data; all we need is the arithmetic mean, the standard deviation and the equation for the normal distribution curve. If we wish, we may calculate a set of ordinates of the curve and sketch in the curve itself. In this chapter we consider how to test whether the normal distribution curve is a satisfactory model of the observed distribution. The word "satisfactory" implies some decision about an acceptable level of satisfaction and, indeed, how satisfaction is measured in the first place: we shall come to that in a moment. In this chapter we shall also introduce another very important distribution in statistics, the Chi-square distribution and an important concept, "degrees of freedom".

7.2 Comparison of Observed and Modelled Data

Let us imagine that we have a histogram of some data, perhaps the distribution of head breadths of Cambridge undergraduates which we considered in Chapters 5 and 6, or rainfall from year to year, or the heights of a large sample of the population. We learnt in Chapter 6 how to fit the normal distribution curve, that is we learnt how to

Basic Mathematics for Students of Air Pollutants
By Robert Maynard and Richard Atkinson
© Robert Maynard and Richard Atkinson 2024
Published by the Royal Society of Chemistry, www.rsc.org

determine the parameters of the normal curve which we assumed to be a reasonable model of our data. We now ask, how well does the model fit the data? By this we mean, are the predictions we can make from the model consistent with the observed data? The model gives us what we may call *expected* values, the data give the *observed* values.

We might think of subtracting one from the other and adding up the differences. But, as with the definition of the variance (Chapter 5) this is problematic and hence differences are first squared to ensure they are all positive. Similarly, absolute differences do not take account of the magnitude of the individual values and so some form of scaling is required. The most appropriate statistic is therefore:

$$\sum_{i=1}^{k} \frac{\left(o_i - e_i\right)^2}{e_i}$$

in which "*o*" and "*e*" refer to observed and expected values, *k* the number of observations, *i* stands for the *i*th values of "*o*" and "*e*" and the sigma symbol (Σ) means sum over all values of *i*.

We can see, immediately, that the closer the observed values correspond with the expected values then the lower the value of the total. So, we have a means of quantifying the total difference between observed and expected values. We now face the problem of working out an objective method of deciding when the total difference is small enough to accept the premise, or hypothesis, that the normal distribution is a good model for our data (or, to put it another way, the data we have observed follow a normal distribution). Sounds difficult. But help is at hand in the form of the Chi-square distribution.

7.3 The Chi-square Distribution

The Chi-square distribution is an important distribution in statistics. It is used frequently in hypothesis testing and in the estimation of variance. Any variable that can be written as a sum of squares of independent standard normal variables has a Chi-square distribution. This formal definition can be written thus:

$$\chi^2 = \sum_{i=i}^{v} Z_i^2$$

where χ^2 (χ, a rather curly X, is the lower case version of the Greek letter Chi), Z_i is the standard normal deviate we have met before (see

Chapter 6) and "v" is the number of variables. If we have a standard normal variable, *i.e.* mean 0 and variance 1 then if we square each value and then add them up, the sum follows the Chi-square distribution. We can also add more standard normal variables (up to "v") and their sum also follows a Chi-square distribution. "v" is known as the degrees of freedom of the corresponding Chi-square distribution. We will say more about "v" shortly.

As with the normal distribution, the mathematical description of the χ^2 distribution curve is somewhat complicated. The equation is given below but we will omit details of its derivation.

$$y = c\left(\chi^2\right)^{\frac{v-2}{2}} e^{-\frac{\chi^2}{2}}$$

where c is a constant which depends on v and, in Alder and Roessler's words (1), "is determined in such a way that the total area under the probability curve is equal to 1". We shall not say any more about c.

How does the Chi-square distribution help us in our assessment of the goodness of fit of the normal distribution to our data? The answer lies in the fact that the total described in eqn (7.1) follows a chi square distribution, *i.e.*

$$\chi^2 = \sum_{i=1}^{k} \frac{\left(o_i - e_i\right)^2}{e_i} \tag{7.1}$$

where χ^2 is the Chi-square statistic. We know the distribution of this statistic – the Chi-square distribution – and hence we can find the probability of obtaining a value of the statistic or one at least as big assuming our hypothesis that the distribution from which the observed data came is true (here we hypothesised the data came from a normal distribution, but other distributions can also be used).

7.4   Degrees of Freedom

In Section 7.3 we mentioned "degrees of freedom". It is an important parameter of the Chi-square distribution as it determines the shape of the Chi-square distribution curve. We know that the normal curve has but one shape: always bell shaped and always symmetrical about the arithmetic mean. The χ^2 distribution is very different from the normal

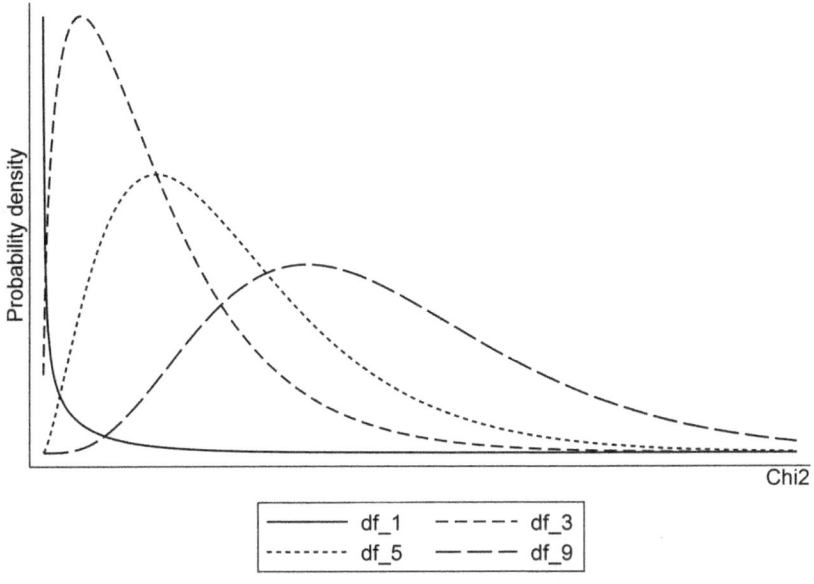

Figure 7.1 Chi-square distribution for 1, 3, 5, and 9 degrees of freedom.

distribution when the number of standard normal variables as indicated by "v" is small; when v is large the χ^2 come to resemble the normal curve. Figure 7.1 shows this clearly.

Everybody who begins to study statistics very soon runs into the phrase "degrees of freedom" (df). We have introduced df as part of the formal definition of the Chi-square distribution, but the term pops up in other areas of statistics too: in the calculation of simple descriptive statistics, in statistical tests and regression analysis. But what is meant by "degrees of freedom"? Most textbooks of statistics provide explanations of the phrase but, in some mysterious way, these tend to differ.

We will not attempt a formal definition. Instead, we aim to provide the student with an intuitive sense of what df are by use of some simple examples.

First, imagine we have a set of 10 measurements from $PM_{2.5}$ monitors recording daily average concentrations ($\mu g\ m^{-3}$) across a city. The data on any particular day may look something like this:

Day "i": 12.1, 16.3, 9.2, 8.2, 11.1, 21.0, 12.1, 12.8, 13.7, 15.5

The average of these values is the sum divided by 10, 13.2. Now suppose one of these values was lost from the data but we still knew the mean:

Day "i": 12.1, 16.3, 9.2, 8.2, 11.1, 21.0, 12.1, 12.8, 13.7, ?

With this information we could, of course, calculate quite simply the missing value. From the mean we can find the sum = 132 and the missing values is then $= 132 - (12.1 + 16.3 + ... + 13.7) = 15.5$ as expected (of course if we knew only the sum then the calculation would be even easier).

In calculating the mean we have imposed a constraint on the data. For the given mean, only 9 data points are free to vary, with the 10th fixed because of the imposed constraint (the mean). The last value and the mean are completely dependent on each other. The degrees of freedom in this system is therefore 9, $10 - 1$ or, more generally, $n - 1$ for "n" data points. We came across this idea in Chapter 5 when we gave the formula for the sample variance (and sample standard deviation). As the mean is calculated from the sample data imposing a constraint on the data, $n - 1$ is used rather than n.

More generally, the df can be defined as $n - k$ where "k" is the number of constraints or parameters estimated from the data.

By way of another example, let us consider a study to assess the association between long-term exposure to air pollutants and chronic respiratory diseases. For each person in the study, we can cross-tabulate the number of people exposed and unexposed to air pollution against disease status. This is illustrated in Table 7.1:

Given that the total number of subjects in the study (T), the total number of diseased/disease free subjects (c_1 and c_2) and exposed/unexposed subjects (a and b) are, once determined, fixed, then only one of the frequencies a, b, c or d are free to vary, since specifying any one of these values enable all of the other values to be derived. In fact, we have two constraints, one for the rows and one for the columns. The df is then the number of rows $- 1 \times$ number of columns $- 1$, $r - 1 \times c - 1$. In this example, df $= (2 - 1) \times (2 - 1) = 1$. If exposure was classified as high, medium or low then the corresponding df would be $(3 - 1) \times (2 - 1) = 2$. It is easy to see how this generalises to an $r \times c$ table.

Let us consider our problem when using the χ^2 distribution to examine the fit of a normal distribution to data displayed as a histogram. Our problem lies in using the table of χ^2. Let us say we have k classes (k blocks in the histogram). Three numbers are fixed: the number of classes, the mean and the standard deviation. Thus, the number of degrees of freedom $= k - 3$. Now then, we have the calculated value of χ^2 and we know the number of degrees of freedom. With these two

Table 7.1 Schematic of a 2 × 2 contingency table reporting exposure status against disease outcome.

Exposure/disease status	No disease	Disease	Totals
Exposed	a	b	r_1
Unexposed	c	d	r_2
Totals	c_1	c_2	Overall total (T)

numbers we can consult the table of χ^2. The table gives the area of the tail of the distribution at various values of probability, p. Assuming we are using the $p = 0.05$ (1 in 20) level of probability as our test we read across the table for the row labelled $\nu = 4$ (the calculated number of degrees of freedom). The value in the column headed A (area of the tail) $= 0.05$ is the value we want. Let us take an example and see how it works out in practice.

7.5 An Example of the Use of the Chi-squared Test

For an example let us return to our 1000 Cambridge students and the breadths of their heads. The mean and standard deviation of these measurements were 6.06″ and 0.2012″ respectively. Using the normal distribution curve with these values for the mean and standard deviation, areas under the normal curve for the ordinates specified by the class boundaries were calculated. Table 7.2 sets out the class

Table 7.2 Head breaths of Cambridge students: observed frequencies (*o*), expected frequencies (*e*) and calculation of the Chi-square statistic[a]

	Class boundaries	*o*	*e*	*o – e*	$(o-e)^2$	$(o-e)^2/e$
5.5	[5.45, 5.55)	3	4.3	−1.3	1.8	0.4
5.6	[5.55, 5.65)	12	14.9	−2.9	8.5	0.6
5.7	[5.65, 5.75)	43	40.4	2.6	6.9	0.2
5.8	[5.75, 5.85)	80	85.8	−5.8	33.7	0.4
5.9	[5.85, 5.95)	131	143.1	−12.1	147.3	1.0
6.0	[5.95, 6.05)	236	187.5	48.5	2355.7	12.6
6.1	[6.05, 6.15)	185	192.8	−7.8	60.1	0.3
6.2	[6.15, 6.25)	142	155.6	−13.6	185.0	1.2
6.3	[6.25, 6.35)	99	98.6	0.4	0.2	0.0
6.4	[6.35, 6.45)	37	49.1	−12.1	145.5	3.0
6.5	[6.45, 6.55)	15	19.2	−4.2	17.3	0.9
6.6	[6.55, 6.65)	12	5.9	6.1	37.5	6.4
6.7	[6.65, 6.75)	3	1.4	1.6	2.5	1.8
6.8	[6.75, 6.85)	2	0.3	1.7	3.0	11.3
Totals		1000	998.8			39.9

[a]Note the class boundaries. In mathematical notation, an interval bounded by a curved bracket indicates that the corresponding limit is not included in the interval whereas a square bracket indicates that the interval includes the limit value. Their use makes clear which intervals contain the boundary values. Note also that the sum of the expected values is not 1000. This is because we are estimating the expected frequencies from a normal distribution. One further point to mention is that the normal distribution curve is a reasonable predictor of frequencies only when the number of observations per class is >5. One might therefore consider combining some categories for a more careful assessment.

boundaries used as the ordinates. It also shows the observed values (o) and the values expected from application of the normal curve (e) and the calculation of the Chi-square statistic.

The calculation of the χ^2 statistic is a simple matter of carrying out the necessary arithmetic as shown in Table 7.2. The method of calculating the area under the normal curve between two values (the expected frequencies) was shown in Chapter 6 and we shall not repeat that here. The result of our labours is a value for our Chi-square statistic, χ^2 = 39.9. Now we need the table of Chi-squared: see Table A3 in the Appendix.

We have 14 classes and, as indicated earlier, 3 parameters. Hence, 14 − 3 = 11 df. We look down the column headed k (for degrees of freedom) and find the row labelled 11 and read across the columns. The last column headed 0.001 gives a χ^2 value of 31.264. Note that the table gives us the area to the right of the ordinate which cuts off, so to speak, the areas appearing at the heads of the columns: 0.99, 0.98···0.20, 0.10, 0.05, 0.02···0.001.

Now we compare the value we calculated for χ^2, that is 39.9, with 31.2. It is larger! This means the probability of obtaining a value at least as big as 39.9 is less than 0.001. A computer software package can give us the exact P-value: 0.000037. Very unlikely! So, we conclude that we have evidence to reject the hypothesis that the normal curve is a reasonable representation of the data. In Chapter 5 we introduced the head breadth data as an example of how the distribution of real data can follow, though not exactly, that of the normal distribution. Our formal statistical test of this hypothesis suggests these data are not normally distributed. However, the normal approximation may be good enough to support analyses, for example using the mean as a measure of central location. Other descriptive and statistical tests to assess normality are available. These include Q–Q plots, measures of skewness and kurtosis, Shapiro–Wilks and Kolmogorov–Smirnov tests (see Further Reading).

We have shown how to calculate the χ^2 statistic from first principles and how to derive the appropriate P-value from a table of the χ^2 distribution. Statistical software can perform all of this analysis easily and quickly. But there is no substitute for undertaking at least one set of calculations by hand in order to understand fully the theory and interpretation of the analyses. We encourage the reader to do so.

The Chi-squared distribution is a very useful distribution in that it allows us to test the fit between an observed distribution of values and an assumed parametric distribution and, also, to test whether the pattern of counts of defined events conform to what would have been expected had the pattern been dictated by chance alone. For further details, see Further Reading.

8 Distribution of the Diameters of Particles in a Typical Aerosol

8.1 Introduction

Much of what we have learnt from previous chapters can be applied to an analysis of the distribution of diameters of particles in a typical aerosol. In this chapter we shall use an example provided by Hinds (see Further Reading) and explore the several ways in which the distribution of particle diameters may be displayed. The emphasis of this chapter is on graphical representations rather than on mathematical analysis; the graphs present the essential information. But we shall have to say something about the mathematics especially as regards the use of the geometric mean (which we shall see is identical to the median) and the geometric standard deviation of the data around the geometric mean. We have met the geometric mean, we shall soon meet the geometric standard deviation.

8.2 Using Histograms to Understand Data

Let us begin by examining the table of data provided by Hinds[1] and reproduced with permission; Table 8.1 contains an expansion of the table provided by Hinds. We have numbered the columns. Column 9 deals with the cumulative (Cum) percentage.

Basic Mathematics for Students of Air Pollutants
By Robert Maynard and Richard Atkinson
© Robert Maynard and Richard Atkinson 2024
Published by the Royal Society of Chemistry, www.rsc.org

Table 8.1 Diameters of particles from a typical aerosol shown as grouped data. Adapted from ref. 1 with permission from John Wiley & Sons, Copyright © 1999, John Wiley and Sons Inc[a]

(1) Class size range μm	(2) Class width μm	(3) Class midpoint μm	(4) Count (frequency, number)	(5) Frequency per μm	(6) Fraction	(7) Fraction per μm	(8) % (Count as % total)	(9) Cum % (Count as % total)
0–4	4	2	104	104/4 = 26	104/1000 = 0.104	0.104/4 = 0.026	10.4	10.4
4–6	2	5	160	80	0.160	0.08	16.0	26.4
6–8	2	7	161	80.5	0.161	0.0805	16.1	42.5
8–9	1	8.5	75	75	0.075	0.075	7.5	50.0
9–10	1	9.5	67	67	0.067	0.067	6.7	56.7
10–14	4	12	186	46.5	0.186	0.0465	18.6	75.3
14–16	2	15	61	30.5	0.061	0.0305	6.1	81.4
16–20	4	18	79	19.7	0.079	0.0197	7.9	89.3
20–35	15	27.5	90	6.0	0.090	0.0060	9.0	98.3
35–50	15	42.5	17	1.133	0.017	0.0011	1.7	100.0
>50			0					100.0
Total			1000				100.0	

[a]Note: Intervals are equal to or greater than the lower limit and less than the upper limit.

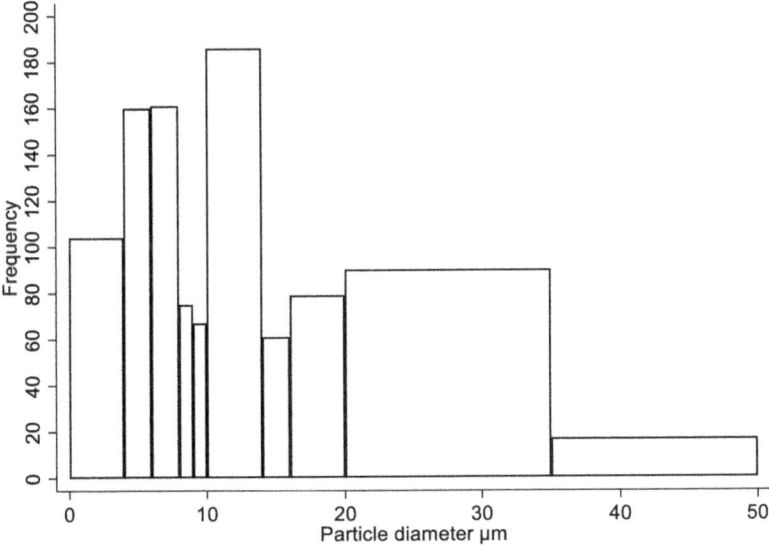

Figure 8.1   Histogram of particle diameters in Table 8.1.

Examination of the table shows that the size intervals shown in the first column are not of equal width. This contradicts the advice given in textbooks of elementary statistics. As suggested in Chapter 5 a histogram of the data can be plotted: Figure 8.1.

Note, as always, the labelling of the axes: particle diameter in microns on the *x* axis, and frequency (or count) on the *y* axis. Recalling our earlier discussion, we might prefer to label the *y* axis as count (or number) to the nearest micron per class width as shown. This is a little cumbersome. Note the relatively large area of the block for the interval 20–35 μm. Examination of the table suggests that the height of this block is disproportionate to its importance: only 9% of the particles fall into the size range defined by this block and yet it seems to dominate the picture. The reason, of course, is that the height of a block depends on its width, *i.e.* it is the area of each block that is important in a histogram. This may not be immediately obvious. Consider Figure 8.2.

Here we have a histogram in which the blocks are of equal width, 1 unit. We can see that the number of observations (10), shown in the *y* axis, is identical from block to block. Hence, in each block we have 10 × 1 observations: 60 observations in total. Figure 8.3A shows what happens if we combine three of the blocks. The "triple block" now contains 30 observations, and its height has increased by a factor of 3.

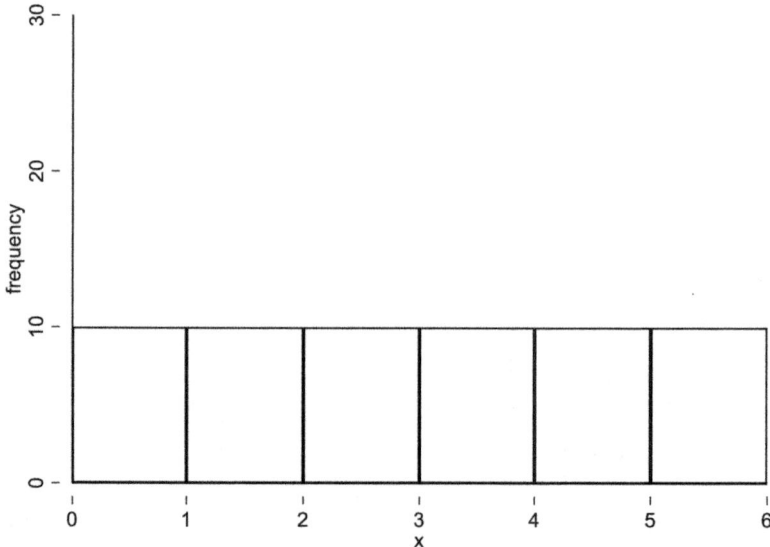

Figure 8.2 Histogram with identical frequencies for equal intervals of variable *X*.

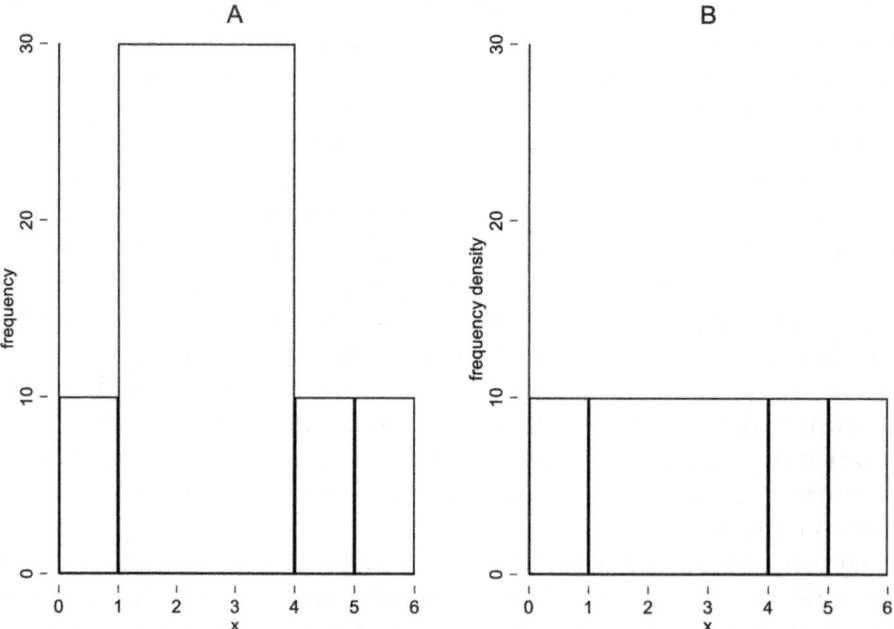

Figure 8.3 Histogram with identical frequencies for equal intervals of variable *X*. In Figure 8.3A the *y* axis displays frequency and in 8.3B
frequency density.

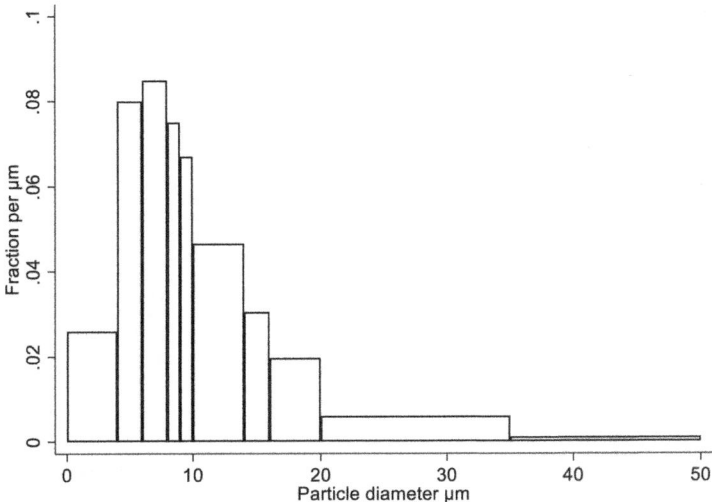

Figure 8.4 Distribution of particle diameters with the *y* axis representing fraction per μm particle diameter.

This is misleading. Why? In a histogram the value of any interval is represented by the area and not the height of the block, as in a bar chart. Recalculating the total number of observations would give us 10 + (30 × 3 = 90) + 10 + 10 = 120. This undesirable effect of variability in the widths of intervals is easily overcome by plotting the number per unit diameter: we divide the frequencies by the widths of the classes. The effect of this simple modification is shown in Figure 8.3B. We can see that *the height of* the "1–4" block now reflects its actual importance.

Returning to our example, we could plot the frequency of particles per unit diameter or the fraction of the total number of particles per unit diameter, precisely the same distribution as shown in Figure 8.4, would be produced but the *y* axis is now labelled fraction per μm (not frequency) and, of course, the numbers on the *y* axis have changed.

In Figure 8.4 the total number of particles in each block is determined by multiplying the height of the block by 1000 by its width. So, for interval 20–35 μm, 0.006 × 1000 × 15 = 90 particles in size range 20–35 μm.

It is easy enough to turn Figure 8.4 into an approximately smooth curve by joining the *mid-points* of the tops of the blocks: Figure 8.5 shows the curve.

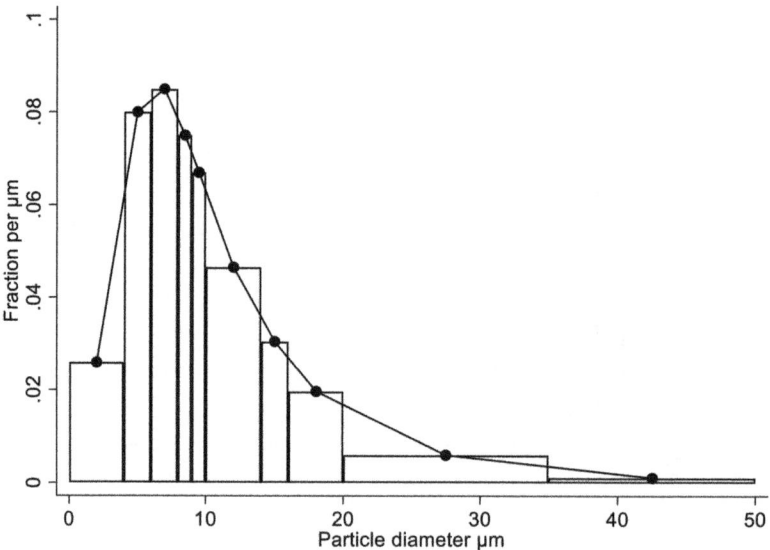

Figure 8.5 Distribution of particle diameters with the *y* axis representing
fraction per μm particle diameter with frequencies connected
to illustrate the distribution curve.

8.3   Skewed Data and the Log-normal
Distribution

Examination of the curve shows that it is skewed: the peak is near the
y axis and there is a long tail to the distribution which runs away to
the right. This is a positively skewed distribution – the tail extends in
the increasingly positive direction (as opposed to a negatively skewed
distribution where the tail extends towards negative values). Examine
the plot closely: confirm that the points plotted really are the mid-
points of the size ranges, that is, the mid-points of the tops of the
blocks in Figure 8.5. We recall from Chapter 3 that some curves can be
turned into straight lines by plotting the logarithms of one of the variables; we shall now see that a skewed distribution such as that shown
in Figure 8.5 can be turned into a normal distribution curve by taking
logs of the diameters of the particles and plotting fraction per unit
log particle diameter against log particle diameter. Fraction per unit
log particle diameter is abbreviated to fraction/$\Delta \ln d$. A skewed distribution which, when treated in this way, yields a normal distribution
is, very reasonably, called a log-normal distribution (see later in this
chapter for a mathematical treatment). We shall need to make some

Table 8.2 Essential calculations to determine fraction per unit log particle diameter.

Diameter (d) μm	Natural logarithms of the range of diameter in each class	Class width in log units	Fraction	Fraction/Δln d
0–4	[a]	[a]	0.104	(0.05)[a]
4–6	1.3863, 1.7918	0.4055	0.160	0.3946
6–8	1.7918, 2.0794	0.2876	0.161	0.5598
8–9	2.0794, 2.1972	0.1178	0.075	0.6367
9–10	2.1972, 2.3026	0.1054	0.067	0.6357
10–14	2.3026, 2.6391	0.3365	0.186	0.5527
14–16	2.6391, 2.7726	0.1335	0.061	0.4569
16–20	2.7726, 2.9957	0.2231	0.079	0.3541
20–35	2.9957, 3.5553	0.5596	0.090	0.1608
35–50	3.5553, 3.9120	0.3567	0.017	0.0474
>50				

[a]Note the first class has zero μm as its lower bound and 4 μm as its upper bound. It is not possible to subtract ln0 from ln4, given that ln0 is undeterminable: it is exactly equivalent to dividing 4 by 0. A close approximation is to add a small value to the lower limit.

calculations based on the data provided in Table 8.1 before plotting the normal curve of the modified data. The modification, by the way, is called a "transform", in this case a logarithmic transform. Table 8.2 shows the essential calculations.

It will be seen that the boundaries of the various classes have been presented as the natural logarithms of the original diameters. The widths of the classes are calculated as the difference between the upper and lower bounds of the classes. We have subtracted logarithms; we know that this means we have divided the upper bound by the lower bound. If we did that and then took the natural logarithm of the answer, we would arrive at precisely the same figure as produced by subtracting the natural logarithms as shown in the table. We can now calculate fraction per unit $\ln d$ for each class (fraction/$\Delta\ln d$) by dividing the fraction represented by the class by the width of the class expressed in log units.

We could plot the data with fraction/$\Delta\ln d$ on the y axis and $\ln d$ on the x axis, but it is easier to use semi-log graph paper and plot particle diameter mid-point on the x axis. See Figure 8.6.

Examination of Figure 8.6 suggests, pleasingly, that the application of the log transform has converted the skewed distribution into a normal shaped distribution curve. It is hardly necessary to point out that the scale on the y axis has changed from that used in Figure 8.5. Having arrived at a normal curve let us do some algebra. In Chapter 6 we

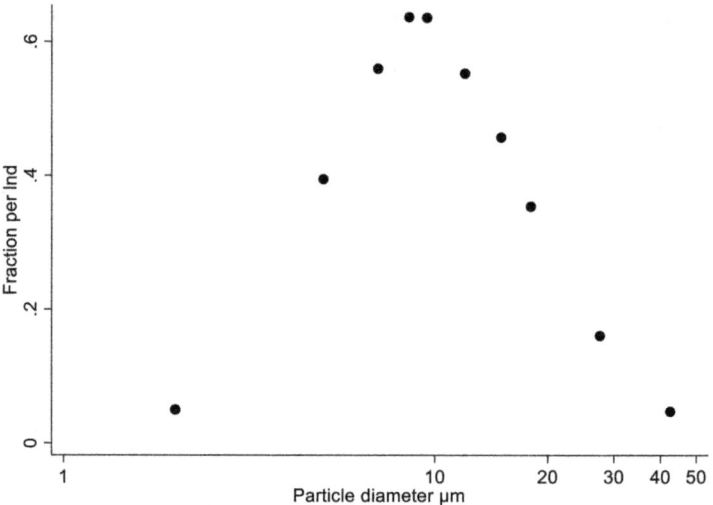

Figure 8.6 Fraction/$\Delta \ln d$ against mid-point natural logarithm particle diameter.

exhausted ourselves in considering the equation of the normal distribution curve; let us remind ourselves of it.

$$y = \frac{1}{\sigma\sqrt{2\pi}} e^{-\frac{1}{2\sigma^2}(x-\mu)^2}$$

When considering a log-normal distribution we must remember that we will be working with logarithms of x (natural logarithms are used), the geometric mean, and the geometric standard deviation.

Let us take these in turn. We need not say anything about finding the natural logarithms of values of x: a calculator will give the natural logarithm of any positive number we choose to consider.

We considered the geometric mean in Chapter 5, let us remind ourselves of the equation

$$\log m_g = \frac{\sum_{i=1}^{n} \log x_i}{n}$$

Or, using natural logarithms

$$\ln m_g = \frac{\sum_{i=1}^{n} \ln x_i}{n}$$

The equation for the ordinary standard deviation of a set of data was also considered in Chapter 5:

$$s = \sqrt{\frac{\sum_{i=1}^{n}(x_i - m)^2}{n-1}}$$

We could, of course, write σ for the population SD and use μ instead of m and N instead of $n - 1$.

Now then, the natural logarithm of the logarithmic standard deviation (the geometric standard deviation) for a set of data will be

$$\ln s_g = \sqrt{\frac{\sum_{i=1}^{n}(\ln x_i - \ln m_g)^2}{n-1}}$$

To calculate $\ln s_g$ we must have to hand, so to speak, the natural logarithms of the values of x and the natural logarithm of the geometric mean. Let us work through a short series of data. We shall use m for the arithmetic mean, m_g for the geometric mean, s for the ordinary (arithmetic) standard deviation and s_g for the geometric standard deviation. Σx has been used for $\sum_{i=1}^{n} x$, and similarly for all the columns of the table. The essential calculations to determine geometric mean and geometric standard deviation are given in Table 8.3.

Working out totals: $\Sigma x = 62$, $\Sigma(x - m)^2 = 595.36$, $\Sigma \ln x = 10.3971$, $\Sigma(\ln x - \ln m_g)^2 = 4.8044$

Applying the formulae given above: $m = 12.4$, $s = 12.2$, $m_g = 8$, $s_g = 3$

The attentive reader will have noticed that the geometric mean is identical to the median of the data set. This is inevitably the case; taking the logarithms of the observations does not alter the order in which they appear. We shall see that the term "median" is substituted for the term "geometric mean" in expressions such as "count median diameter" of the particles in an aerosol.

Table 8.3 Essential calculations to determine geometric mean and geometric standard deviation.

x	$x - m$	$(x - m)^2$	$\ln x$	$\ln x - \ln m_g$	$(\ln x - \ln m_g)^2$
2	−10.4	108.2	0.6931	1.3863	1.9218
4	−8.4	70.6	1.3863	0.6931	0.4804
8	−4.4	19.4	2.0794	0	0
16	3.6	13.0	2.7726	0.6931	0.4804
32	19.6	384.16	3.4657	1.3863	1.9218

The equation of the log-normal distribution is

$$y = \frac{1}{\ln \sigma_g \sqrt{2\pi}} e^{-\frac{1}{2(\ln \sigma_g)^2}(\ln x - \ln \mu_g)^2}$$

Note that $(\ln \sigma_g)^2$ is not the same as $\ln \sigma_g{}^2$.

In the first case we are squaring the natural logarithm of σ_g; in the second case we are taking the natural logarithm of the square of σ_g. In case this is not obvious, let $\sigma_g = 3$, and compare

$$\left(\ln \sigma_g\right)^2 = 1.0986^2 = 1.2069$$

$$\ln \sigma_g{}^2 = \ln 9 = 2.1972$$

$(\ln \sigma_g)^2$ is sometimes, slightly confusingly, written as $\ln^2 \sigma_g$.

8.4   Cumulative Plots of the Log-normal Distribution

If the cumulative frequency is plotted against $\ln x$ an ogive will be produced. Just as in the case of the ordinary normal distribution this ogive can be converted into a straight line by using probability paper, but, of course, we shall need to use log-probability paper with the x axis arranged logarithmically and the y axis arranged in conformity with the distribution of the normal curve. Figure 8.7 shows a cumulative curve (*not* based on the data in Table 8.1) plotted on log-probability paper.

Note the scale on the y axis. We have marked in horizontal lines at 84% and 16%. Why? Because these percentages represent the upper and lower boundaries of 68% of the data. It will be recalled from our discussion of the ordinary normal curve that the central 68% of the data is bounded by +1 and −1 standard deviation either side of the mean. Now we must be careful. Recall that we have plotted values of x on a logarithmic scale. We could, of course, have produced the same graph by plotting the natural logarithms (or, indeed, the common logarithms) of x on the x axis. Let us read off the values of x for the median (geometric mean) = 1.10; the value of x corresponding to 84% on the y axis = x_b = 2.1, and the value of x corresponding to 16% on the y axis = x_a = 0.564. Had we been plotting the natural logarithms of x we

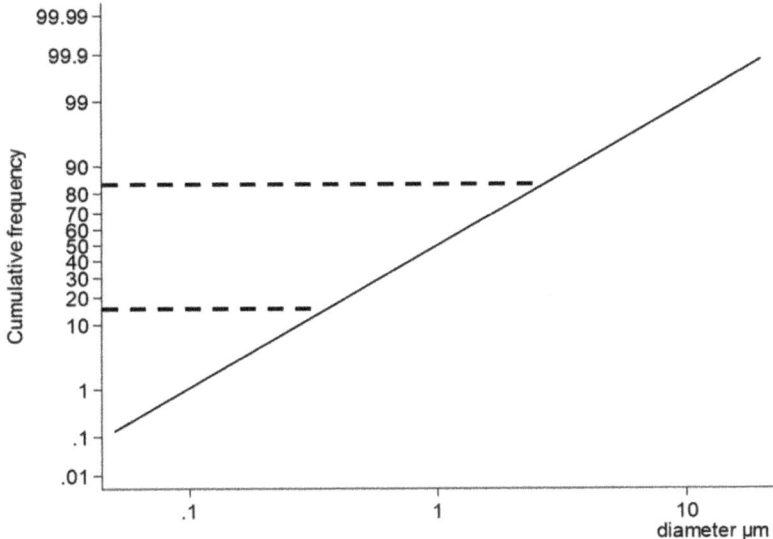

Figure 8.7 Plot of cumulative frequency against natural logarithm of particle diameter.

would have argued that $\ln 2.1 - \ln 1.10 = \ln s_g$. We know that we could have written this as $2.1/1.10 = s_g$. We can, then, calculate s_g very simply by dividing the 84% value of x by the 50% value of x, always remembering that we have plotted values of x on a log scale. Put in algebraic terms the population geometric standard deviation is thus

$$e^{\ln x_b - \ln \mu_g} \text{ or } e^{\ln \mu_g - \ln x_a} \text{ or } \frac{x_b}{\mu_g} \text{ or } \frac{\mu_g}{x_a}.$$

To repeat, the geometric standard deviation is the 84% size *divided by* the 50% size and the 50% size *divided by* the 16% size. The enthusiast will have appreciated that

$$\sigma_g = \frac{x_b}{\mu_g} = \frac{\mu_g}{x_a}$$

$$\sigma_g^2 = \frac{x_b}{\mu_g} \cdot \frac{\mu_g}{x_a} = \frac{x_b}{x_a}$$

$$\sigma_g = \sqrt{\frac{x_b}{x_a}}$$

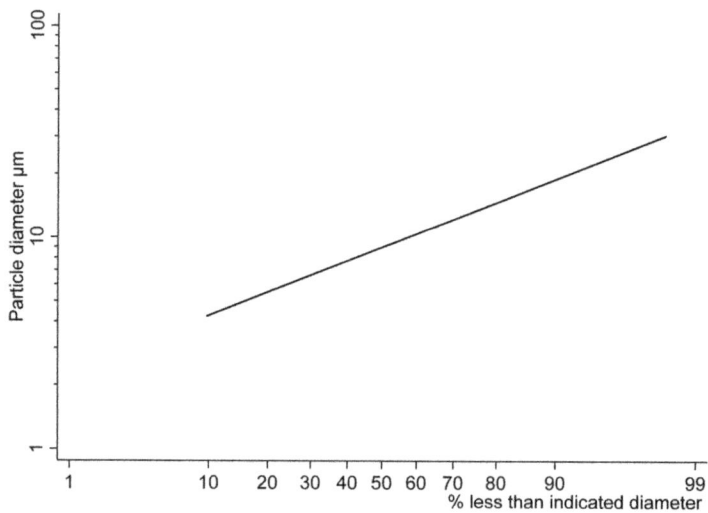

Figure 8.8   Log-probability plot of the data presented in Table 8.1.

Let us return to the data introduced in Table 8.1 above. Figure 8.8 shows the log-probability plot of the data presented in Table 8.1 *with the axes in the conventional positions*: diameter on the vertical axis, percentages on the horizontal axis.

  This is the most important graph in this chapter. It is much easier to plot than the normal distribution curve shown in Figure 8.7; we do not need to make any additional calculations, all the data we need are provided in Table 8.1. In practice there is no need to plot the normal curve as shown in Figure 8.7, we can move from the data to the cumulative log-probability curve. It will be obvious that if the cumulative log probability curve is not a straight line (or close to one) then the data do not conform to a log-normal distribution.

  Note that we may read off the geometric mean (the median) and the 84% diameter and the 16% diameter. Dividing the 84% size by the 50% size gives the geometric standard deviation: 1.89. The geometric mean (median) or, we should say count median diameter (CMD) is 9.0 μm. Note, finally, that the scale on the horizontal axis is labelled percentage *less* than indicated size.

References

1. W. Hinds, *Aerosol Technology*, John Wiley and Sons Inc, New York, 2nd edn, 1999.

9 The Statistical Distribution of Mass and Surface Area of Particles Comprising an Aerosol

9.1 Introduction

We have discovered how to characterise a log-normal distribution of the diameters of particles in a sample aerosol in terms of count median diameter (CMD) and geometric standard deviation (GSD). We now turn to other parameters which can be used to describe the aerosol. Many parameters have been identified and, very interestingly and very helpfully, they are all related to the count median diameter by the geometric standard deviation and a series of equations known as the Hatch–Choate equations. We shall not attempt to derive these equations; the rather advanced mathematics of the derivations is given by Hinds (see Further Reading).

9.2 The Distribution of Number, Mass and Surface Area

We have seen that if we plot the diameter of the particles on a logarithmic scale along the x axis and the fraction of the total number of particles per Δln diameter on the y axis we obtain a normal distribution

Basic Mathematics for Students of Air Pollutants
By Robert Maynard and Richard Atkinson
© Robert Maynard and Richard Atkinson 2024
Published by the Royal Society of Chemistry, www.rsc.org

curve. We noted that the curve could be described by its median (count median diameter) and geometric standard deviation which describes the spread of the curve. What may be a surprise to us is that if we plotted the fraction of the total surface area of the particles per Δln diameter or fraction of the total mass of the particles per Δln diameter, on the *y* axis and, as before, particle diameter on the *x* axis on a logarithmic scale, we would obtain further normal curves and, most importantly, all these curves would have the same geometric standard deviation. A set of curves is shown in Figure 9.1.

You will see that as we move from count to surface area to mass the curves are displaced to the right along the *x* axis. Jumping ahead, we can imagine that the equivalent log-probability cumulative distribution curves (actually straight lines) also form a series but stacked one above another: count at the bottom, area above count, mass above area. Because geometric standard deviation controls the slope of these lines and the geometric standard deviation is constant, the lines will be parallel to one another.

A series of equations has been derived to allow us to inter-convert count, surface area and mass median diameters. These are just a few of the series of the equations already referred to as the Hatch–Choate equations. These equations allow the determination of other parameters of the aerosol including diameter of average mass and mass mean diameter (these are not the same thing: see Hinds (Further Reading)

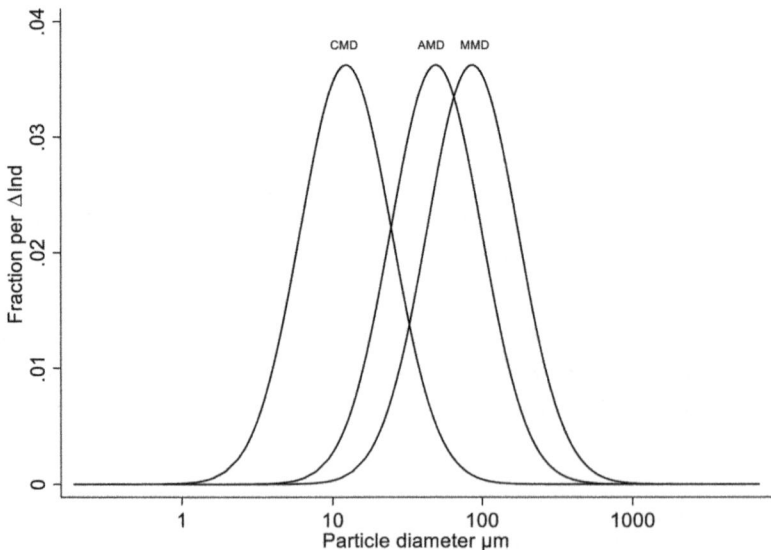

Figure 9.1 Distributions of particle count median diameter (CMD), area median diameter (AMD) and mass median diameter (MMD).

for details). For our purposes we shall consider only surface area median diameter and mass median diameter.

Which Hatch–Choate equations concern us? They are, using AMD and MMD for area and mass median diameter respectively,

$$AMD = CMD \exp\left(2\ln^2 \sigma_g\right)$$

$$MMD = CMD \exp\left(3\ln^2 \sigma_g\right)$$

Recall, again, that $\ln^2 \sigma_g = \left(\ln \sigma_g\right)^2$ and NOT $\ln \sigma_g^2$.

Let us do the algebra. Given $CMD = 9\mu m, \sigma_g = 2$

$$AMD = 9 \times \exp\left(2 \times 0.7 \times 0.7\right)$$

$$AMD = 9 \times \exp 0.98$$

$$AMD = 9 \times 2.7 = 24.3$$

$$MMD = 9 \times \exp\left(3 \times 0.7 \times 0.7\right)$$

$$MMD = 9 \times \exp 1.47$$

$$MMD = 9 \times 4.349 = 39.14$$

Of course we could write, for example,

$$\ln MMD = \ln CMD + 3\ln^2 \sigma_g$$

$$\ln MMD = 2.197 + 1.47 = 3.667$$

$$MMD = 39.14$$

Let us look at the log-probability plots: Figure 9.2. Remember, the lines are parallel; GSD is a constant.

Let us examine the plots. Note that the lines representing the distribution of particle number (count), surface area and mass are parallel; that the count, area and mass median diameters are very different, that for each curve we can read off the percentage of the distribution occupied by particles of *less* than a specified diameter; and that we can for each curve read off the percentage of the distribution lying between two specified diameters.

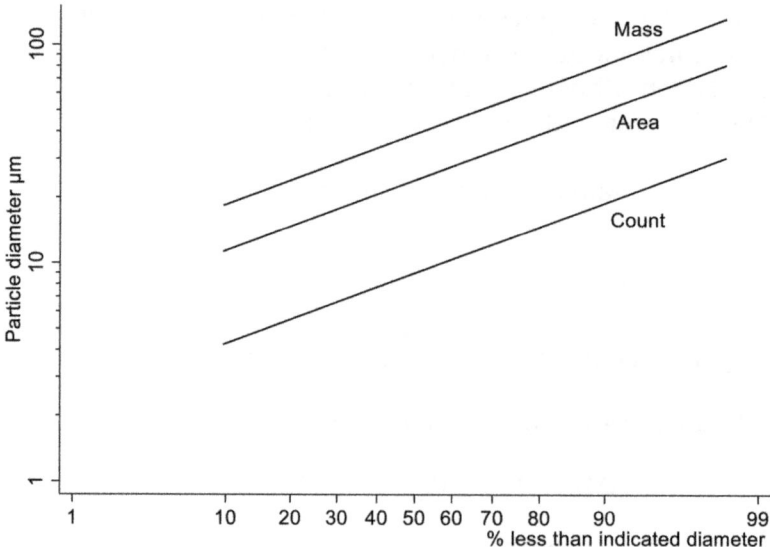

Figure 9.2   Relationships between particle count median diameter (CMD),
area median diameter (AMD) and mass median diameter (MMD)
and percentage of particles less that indicated diameter.

10 Deposition of Particles in the Respiratory Tract

10.1 Introduction

In this chapter we shall consider the deposition of airborne particles in the respiratory tract and outline how an understanding of deposition processes has influenced the metrics or descriptors of ambient concentrations of particulate matter. We shall consider, but not derive, some laws of physics that describe the movement of particles in air; only elementary algebra will be used.

Inhalation of ambient air carries airborne particles into the respiratory system where a fraction of these particles is deposited. In very round figures about half of all the inhaled particles are deposited. Before going any further, we should stress that most of the particles deposited in the respiratory system are removed, or cleared, from the system by the defence mechanisms of the lung. These include the upward movement of mucus by the ciliated cells of the airways and the uptake of particles by macrophages which patrol the inner surfaces of the alveoli and, when loaded with particles, migrate to the airways and step, so to speak, onto the ciliary escalator. Material cleared from the airways may be swallowed or expectorated. A small fraction of the deposited material finds its way into the tissues of the lung: the lungs of residents of polluted cities are, for this reason, slate grey rather than pink in colour. A small fraction, about 1%, of particles of less than 100 μm diameter, ultrafine particles or nanoparticles, finds its way into the blood stream and may be deposited, secondarily, in

Basic Mathematics for Students of Air Pollutants
By Robert Maynard and Richard Atkinson
© Robert Maynard and Richard Atkinson 2024
Published by the Royal Society of Chemistry, www.rsc.org

other organs. Particles deposited in the lung cause an inflammatory reaction and this is thought by toxicologists to underlie, at least in part, the effects of ambient particulate matter on health.

10.2   Anatomy of the Lung

Before describing the mechanisms of particle deposition in the lung we should outline the structure of the respiratory system. The respiratory system comprises the upper airways of the nose, naso-pharynx and larynx, the trachea or wind-pipe and its two branches to the right and left lungs which lead, ultimately and by repeated branching, to the alveoli where the essential exchange of oxygen for carbon dioxide takes place. The two branches of the trachea are the main bronchi, left and right. Repeated division of the bronchi leads to about 65 000 very small branches of about 0.5 mm diameter which are called the terminal bronchioles. On average about 16 divisions precede the terminal bronchioles. No gas exchange occurs across the walls of the airways preceding the terminal bronchioles nor does any occur across the walls of the terminal bronchioles. These small airways divide and re-divide although little reduction in diameter now occurs: a series of small airways the walls of which bear pouches (alveoli), follows and, after a number of further divisions, spaces leading only into alveoli are produced. Remarkable though it may seem, the number of alveoli in the human adult has been estimated to be about 300 million. A maximum of about 22 divisions, actually bifurcations, separates the trachea from the alveoli. In the central parts of the lung alveoli may be separated, so to speak, from the trachea by as few as 10 bifurcations. Gas exchange takes place across the walls of the alveoli which, as we have seen, occur distal to the terminal bronchioles; this part of the pathway comprises the gas exchange zone of the lung. All the airways proximal to and including the terminal bronchioles comprise the conducting airways. The alveoli themselves are about 200 µm in diameter. It is important to realise that the great majority of the volume of the lung is provided by the gas exchange zone: the volume of the whole lung of a young adult is about 6 L at full inspiration whereas the volume of the conducting airways is only about 150 cm^3.

Inhaled air flows the through the narrow opening of the larynx into the trachea. Air flow in the trachea and large airways is turbulent but soon settles down, so to speak, to streamline flow in the further airways. The shape of the wave-front of the incoming air is not flat; on the contrary it forms a cone with the apex pointing in the direction of air flow. Bifurcation leads to an increase in total cross-sectional area of the airways and the velocity of flow of air declines progressively; by the time the incoming air reaches the smallest airways bulk flow has just about stopped and oxygen molecules make the last part of their journey to the alveolar walls by diffusion.

Before leaving the anatomy of the lung we should note that aerosol physicists interested in the deposition of particles in the lung refer to the conducting airways as the tracheo-bronchial (TB) compartment of the system and to the gas exchange zone as the pulmonary (P) compartment.

10.3   Particle Deposition in the Lung

Inhaled particles are deposited on the inner surfaces of the airways and alveoli of the lung. The extent to which they are deposited in the various parts of the airways is largely controlled by particle size. The pattern of deposition has been studied by means of (1) mathematical models of the airways, (2) physical models of the airways, and (3) by exposing volunteers to radioactive particles and monitoring both regional deposition and clearance. These studies have produced broadly consistent results, which are illustrated by Figure 10.1.

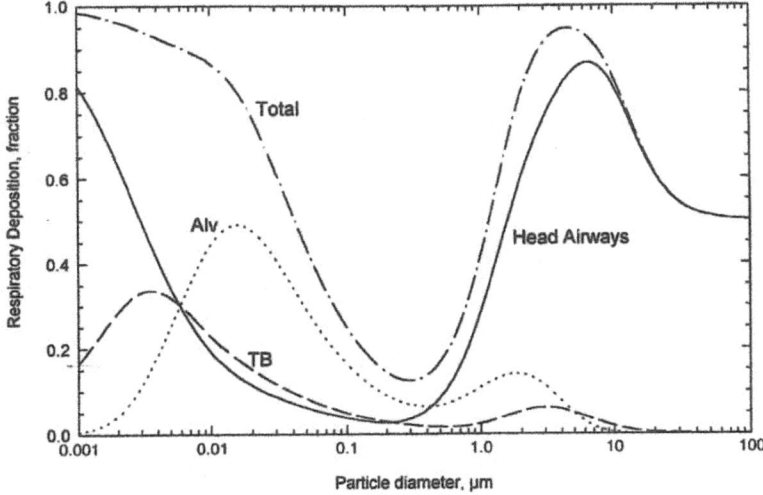

Figure 10.1   Respiratory deposition fraction against particle diameter.[1] Reproduced from ref. 1 with permission from John Wiley & Sons, Copyright © 1999, John Wiley and Sons Inc.

This graph tells us nearly all we need to know about particle deposition in the respiratory tract so let us look at it closely. Note that percentage deposition is shown on the *y* axis: by "fraction" we mean the numerical fraction of particles of the specified sizes. Particle diameter is shown on the *x* axis: the scale on the *x* axis is logarithmic.

Note that for what we shall call "large" particles of >10 μm (10 000 nm) diameter total deposition reaches 100% but deposition in the alveoli is minimal. Why? Because these large particles deposit mainly in the upper airways of the nose and in the larger airways of the lung: we shall discover why they do so very shortly. Note that very small particles of <0.01 μm (10 nm) diameter (<10 nm diameter) follow the same pattern: all that are inhaled are deposited in the respiratory tract, but none, or at least very few, in the alveoli. Again, we shall soon discover why this is so. Note that the total deposition curve looks rather like an upside-down normal distribution curve but note that there are two humps in the alveolar deposition curve. For both total deposition and alveolar deposition there is a point of minimal deposition at about 0.5 μm (500 nm) diameter, and what we shall call the upper peak of alveolar deposition occurs between 1 and 5 μm (1000–5000 nm) diameter. What we shall call the lower peak of alveolar deposition occurs at about 0.02 μm (20 nm) diameter.

Re-examine the graph and make sure that you have all these points clearly in mind. Examine each of the lines with care.

Why is this graph so important?

1. It tells us a lot about deposition of particles in the respiratory tract.
2. It tells us how we should sample ambient particles if we want to discover the fraction of particles likely to be deposited in different parts of the lung.

In fact, the curve dictates the requirements of particle sampling devices. Let us imagine that we could select from the air all inhaled particles of less than, say, 4.0 μm (4000 nm) diameter. From the graph we can see that we would have selected essentially all the particles likely to be deposited in the alveoli. This is the basis of the sampler devised by the UK Medical Research Council (MRC) for sampling particles likely to be deposited in the alveoli: the so-called "respirable dust" sampler. Sampling all the particles of less than 2.5 μm diameter would do almost as well. You may have noticed that we would also collect the very small particles which deposit in the upper airways and not in the alveoli though these would contribute very little to the mass collected. If we sampled all the particles of less than 10 μm (10 000 nm) diameter we would "catch" all the particles likely to be deposited in the airways of the lung.

10.4 Mechanisms of Deposition of Particles

What makes a particle leave the air in which it is suspended and deposit itself on a surface? Several mechanisms are involved in the deposition of particles in the lung; we shall consider them in turn.

10.4.1 Sedimentation

Sedimentation simply means falling under the influence of gravity. If a marble is dropped it falls to the ground under the influence of gravity. We all know that the acceleration of a falling body produced by gravity is 9.81 m s^{-2}. But if a mouse is dropped down the shaft of a coal mine it lands and runs away, or so it is said. Why isn't it smashed to pieces? The answer is that the falling mouse accelerates to a terminal velocity and that velocity is rather low, low enough to allow the mouse to land safely. Of course if a man falls down the shaft he reaches a higher terminal velocity and is killed. Feathers and dandelion seed heads fall in air but rather slowly; they have a very low terminal velocity indeed. It was fortunate for Galileo that he dropped two lead balls from the leaning tower of Pisa; had he dropped a ball of lead and a loosely scrunched up ball of paper he might have been confused. The terminal velocity of a falling body is controlled, in large part, by the resistance imposed by the medium through which the body falls: a marble falls more slowly in treacle than water, and all objects, feathers or balls of lead, continue to accelerate at the same rate when falling in a vacuum.

Stokes showed that the force exerted by gravity on a falling body, F_g, was given by

$$F_g = \frac{4}{3}\pi r^3 \left(\rho_{\text{part}} - \rho_{\text{air}} \right) g \tag{10.1}$$

where r = the radius of the particle, ρ = density (note that the densities of the particle and of air are distinguished) and g = acceleration due to gravity.

He also showed that the resistive force, F_{res}, exerted by air on the falling body was given by

$$F_{\text{res}} = 6\pi\eta r V \tag{10.2}$$

where η = viscosity of air, V = velocity of the body.

For small particles, of the sort of interest to us,

$$F_{res} = 6\pi\eta r \frac{V}{C_c} \tag{10.3}$$

where C_c is a correction factor devised by Cunningham. In essence the correction factor accounts for the tendency for molecules of gas to "slip" past very small falling objects and thus to have a smaller retarding effect than they have on larger objects.

Now then, when the falling particle reaches its terminal velocity, V_t, when $F_g = F_{res}$ and thus from (10.1) and (10.3)

$$6\pi\eta r \frac{V_t}{C_c} = \frac{4}{3}\pi r^3 \left(\rho_{part} - \rho_{air}\right)g$$

Dividing through by r

$$6\pi\eta \frac{V_t}{C_c} = \frac{4}{3}\pi r^2 \left(\rho_{part} - \rho_{air}\right)g$$

Recalling that diameter, d, = $2r$, and that therefore $r^2 = d^2/4$

$$6\pi\eta \frac{V_t}{C_c} = \frac{4}{3}\pi \frac{d^2}{4}\left(\rho_{part} - \rho_{air}\right)g$$

$$6\pi\eta \frac{V_t}{C_c} = \pi \frac{d^2}{3}\left(\rho_{part} - \rho_{air}\right)g$$

Rearranging to give an expression for V_t

$$V_t = \pi \frac{d^2}{3} C_c \frac{1}{6\pi\eta}\left(\rho_{part} - \rho_{air}\right)g$$

$$V_t = \frac{d^2 C_c \left(\rho_{part} - \rho_{air}\right)g}{18\eta}$$

For any specific particle falling in air $\dfrac{C_c \left(\rho_{part} - \rho_{air}\right)g}{18\eta}$ is a constant and thus we can write

$$V_t = Kd^2$$

where

$$K = \frac{C_c \left(\rho_{\text{part}} - \rho_{\text{air}} \right) g}{18\eta}$$

The key point to remember is that V_t depends on d^2. Large particles will have a much higher terminal velocity than small particles and will thus be likely to sediment more readily. This explains the greater likelihood of deposition of large particles than small particles in the large airways of the lung.

10.4.2 Aerodynamic Diameter

Airborne particles come in many sizes and shapes. The sedimentary behaviour of particles varies according to their shape and density. Imagine the seed head of a dandelion: about 3 cm in diameter and likely to float in the air for a lot longer than a 3 cm diameter marble. Imagine an irregular particle, its terminal velocity will be less than that of a spherical particle of the same material and mass. We need some means of standardizing diameter if we are to compare the rates of sedimentation of a range of particles. This standardization is provided by aerodynamic diameter. Let us define aerodynamic diameter.

The aerodynamic diameter of a particle is that of a spherical particle of the same density with a terminal velocity equal to that of the particle of interest.

This is expressed by

$$d_a = d_e \left(\frac{\rho_p}{\rho_o \chi} \right)^{\frac{1}{2}}$$

where, d_a is the aerodynamic diameter, d_e is the equivalent volume diameter, ρ_p is the particle density, ρ_o is the standard particle density (1.0 g cm^{-3}) and χ is a factor that depends on the shape of the particle. For spheres, $\chi = 1$; for cubes, $\chi = 1.08$, for sand particles (varying in shape quite a lot) $\chi = 1.57$; for talc dust, occurring as tiny plates, $\chi = 1.88$. d_e requires definition: we can think of d_e as the diameter of the particle of interest if it were "melted down" into a sphere.

When considering sedimentation of particles of different shapes and densities we need to think in terms of aerodynamic diameter rather than actual diameter. Now then, we remember that when we were discussing the statistical description of aerosols we mentioned mass median diameter and count median diameter.

We can now see that we could have spoken of mass median aerodynamic diameter (MMAD) and count median aerodynamic diameter (CMAD) of an aerosol. This is very important and these terms are, rightly, used when describing aerosols. We need to use these terms if we want to compare the likely sedimentary behaviour of various aerosols.

10.4.3 Impaction

Consider a particle being carried along a tube by flowing air. If the tube, and therefore the line of flow, bends then the particle, by virtue of its inertia, will tend to move across the line of flow and, if in the periphery of the air flow, may impact on the wall of the tube. This is rather like a fast-moving motor car going straight on rather than following the curve of a bend. The distance the particle will travel before continuing in the line of flow of the air can be calculated. The following equation describes the distance travelled by a particle, across the line of flight during negotiation of a bend.

$$D = \frac{v\sin\theta}{g} V_t$$

where D is the distance travelled, v is the velocity before deflection, θ is the angle of deflection of the air-stream and V_t is the terminal velocity as discussed above. We know that terminal velocity is proportional to d^2, so it follows that the likelihood of disposition by impaction is also dependent on d^2.

10.4.4 Diffusion

Molecules of air are in constant motion and collide with aerosol particles. If the particles are small this constant bombardment by molecules causes them to move about in an irregular fashion, this is described as Brownian motion after the microscopist Robert Brown who observed, in 1827, the motion of pollen grains in water. Particles moving in this way may meet the walls of an airway and be deposited. The key point to remember is that only very small particles are significantly deposited by diffusion. The distance (Δ) a particle travels, in three dimensions, by diffusion from its starting point in time, t, is given by

$$\Delta = \sqrt{6Dt} \tag{10.4}$$

where D is the diffusion coefficient of the particle. To be accurate we should say that Δ is actually the root-mean-square displacement: a sort of average for the displacement of a lot of particles of the same size. The equation is sometimes given, for movement along a single axis, as $\Delta = \sqrt{2Dt}$.

D is given by the Stokes–Einstein equation

$$D = \frac{KtC_{\text{c}}}{3\pi\eta d}$$

where T is the absolute temperature, C_{c} is the Cunningham correction factor which we have already come across, K is the Boltzmann constant, η is the gas viscosity, d is the particle diameter. The units of the diffusion coefficient are $m^2\ s^{-1}$.

We may note that for a given temperature and gas, D is proportional to $1/d$.

If we substitute $1/d$ for D in (10.4) then we can see that Δ after a given time will be proportional to $\sqrt{\dfrac{1}{d}}$.

So far so good. Hinds[1] points out that for large particles where the Cunningham correction factor (the slip factor) can be neglected the diffusion coefficient, D, is inversely proportional to particle size: as shown above. But "for small particles with large slip correction factors, D is approximately proportional to d^{-2}." That is, inversely proportional to d^2. This means that for small particles the displacement after a given time will be proportional to $\sqrt{\dfrac{1}{d^2}}$.

This may seem a little confusing but can be explained as follows. Given the following two equations:

$$D = \frac{KtC_{\text{c}}}{3\pi\eta d} \tag{10.5}$$

$$C_{\text{c}} = 1 + \frac{2.52\lambda}{d} \tag{10.6}$$

then, substituting for C_{c} from (10.6) in (10.5) gives

$$D = \left(\frac{1 + \dfrac{2.52\lambda}{d}}{d} \right) \frac{Kt}{3\pi\eta}$$

$$D = \left(\frac{1}{d} + \frac{2.52\lambda}{d^2} \right) \frac{Kt}{3\pi\eta}$$

Let $K_1 = \dfrac{Kt}{3\pi\eta}$ and $K_2 = 2.52\lambda$

$$D = \left(\frac{1}{d} + \frac{K_2}{d^2} \right) K_1$$

Because for all diameters K_1 and K_2 are constants

$$D \approx \frac{1}{d} + \frac{1}{d^2}$$

Note: the symbol "$\approx$" means "approximately equal to".

Let us consider, for a variety of values of d which of the two terms on the right-hand side of the equation is dominant – Table 10.1.

It is clear that as d becomes smaller $\dfrac{1}{d^2}$ becomes the dominant term in (10.4) and as d becomes larger $\dfrac{1}{d}$ becomes the dominant term.

Thus, as particles become smaller, then increasingly,

$$\Delta \approx \sqrt{\left(\frac{1}{d^2} t \right)}$$

Table 10.1 Tabulation of the diffusion coefficient of particles for various values of their diameter[a]

d (μm)	$\dfrac{1}{d}$	$\dfrac{1}{d^2}$	D^*
0.01	100	10 000	10 100
0.1	10	100	110
1	1	1	2
10	0.1	0.01	0.11
100	0.01	0.0001	0.0101

[a]D^*: the values shown in this column are not actual values of D, they show only the effects of varying the relative sizes of $\dfrac{1}{d}$ and $\dfrac{1}{d^2}$. To find the real value of D we would need to take into account the constants, which we have ignored in this explanation. The actual value of D for a spherical particle of 1 μm diameter is 28 μm^2 s^{-1}.

And as particles become larger, then increasingly,

$$\Delta \approx \sqrt{\left(\frac{1}{d}t\right)}$$

It is important to remember that

$$\sqrt{xy} = \sqrt{x}\sqrt{y}$$

Thus as t is a constant, for example 1 s, then as particles become smaller, increasingly

$$\Delta \approx \sqrt{\frac{1}{d^2}}$$

and as particles become larger, increasingly

$$\Delta \approx \sqrt{\frac{1}{d}}$$

We have noted that only a small fraction of very small particles (particles less than, say, 10 nm diameter) are deposited in the alveoli of the lung. Why? The answer is simple: nearly all of these very small particles are deposited *by diffusion* in the upper airways, in fact in the nose.

Let us now consider how far a given particle will travel, by diffusion, in a given period of time. The diffusion coefficient for a given particle is a constant and so we may write

$$\Delta = K\sqrt{t} \tag{10.7}$$

From eqn (10.4) we can see that

$$K = \sqrt{6D} \tag{10.8}$$

Let us express eqn (10.7) in logarithmic terms

$$\log\Delta = \log K + \frac{1}{2}\log t$$

or, to put it in the conventional form for the equation of a straight line

$$\log\Delta = \frac{1}{2}\log t + \log K$$

This straight line has a gradient of ½ and an intercept on the y axis of $\log K$.

Let us consider a particle 1 μm in diameter (Table 10.2). The diffusion coefficient for such a particle at 37 °C in air at 1 atmosphere pressure is 28 μm^2 s^{-1}.

We can calculate from eqn (10.4) that the particle will move, by diffusion, 13 μm in 1 s. Let us calculate how far it will move in 10, 100 and 1000 s.

Recalling from (10.8) that $K = \sqrt{6D}$, then K = the square root of 6×28 = 13. This is of course correct: when $t = 1$ then $\Delta = 13$, Table 10.2.

It is obvious that $\log \Delta$ is increasing linearly with particle diameter.

This can be seen on the following graph (Figure 10.2) where $\log \Delta$ has been plotted against $\log t$.

Some values for the diffusion coefficients of spherical particles are shown in Table 10.3 (data taken, with permission, from Hinds[1]).

Table 10.2 Calculation of Δ and $\log \Delta$ for selected values of t.

T	Δ	$\log \Delta$
1	13	1.1139
10	$13 \cdot \sqrt{10} = 41.6$	1.6191
100	$13 \cdot \sqrt{100} = 130$	2.1139
1000	$13 \cdot \sqrt{1000} = 411$	2.6138

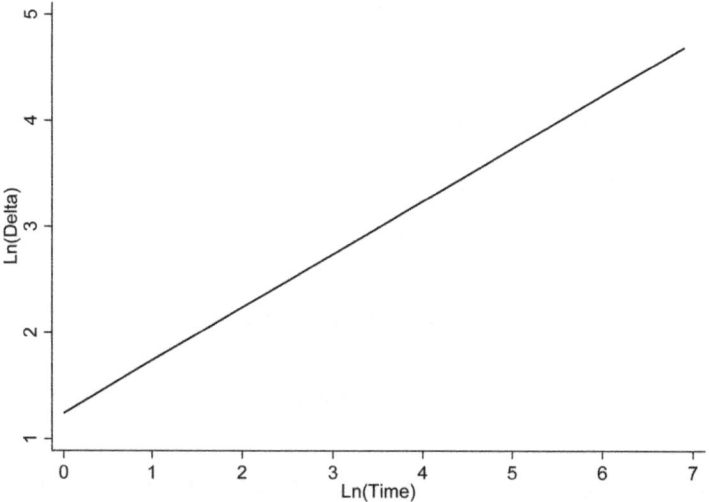

Figure 10.2 Plot of log distance (Δ) a particle travels, in three dimensions, by diffusion from its starting point in time, t.

Table 10.3 Diffusion coefficients of spherical particles of varying diameter.[1] Data from ref. 1.

Particle diameter (μm)	Diffusion coefficient (m^2 s^{-1} and μm^2 s^{-1})[a]
0.00037 (diameter of an "air molecule")	2×10^{-5}, 2×10^{7}
0.01	5.4×10^{-8}, 5.4×10^{4}
0.1	6.9×10^{-10}, 6.9×10^{2}
1.0	2.7×10^{-11}, 2.7×10
10	2.4×10^{-12}, 2.4

[a]1 $m^2 = 1 \times 10^{12}$ μm^2.

It will be obvious that gas molecules move by diffusion much more rapidly than do particles.

The key message is that diffusion will be effective over short distances. One final point. We have spoken of the effective deposition of small particles over short distances. We should also remember that the diffusion of particles is very much slower than that of gas molecules.

10.4.5   Interception

Interception describes the process of deposition of long fibres. Fibres fly like arrows in the air stream and, when a bend comes, tend to go straight on and hit the wall. This is an important mechanism of deposition of, for example, asbestos fibres.

10.4.6   Electrostatic Precipitation

Charged particles are attracted to surfaces which carry opposite charges. Charged particles can induce image charges on uncharged surfaces. All this means that charged particles are more likely to deposit on the walls of the airways than uncharged particles. But is this important? Charge is lost from particles in the air and only in the case of highly charged (recently formed) particles is electrostatic precipitation likely to play an important role in controlling the deposition of particles in the airways.

10.5   Putting it All Together

We can now understand why particles are deposited in the airways. We can see that larger particles are deposited by sedimentation and impaction, and that smaller particles are deposited by diffusion. We can imagine that there is some point, some diameter, when the particles are "too small" to be efficiently deposited by sedimentation and impaction and "too large" to be deposited by diffusion. This point is reached when the particles are about 0.5 μm in diameter. This explains the low point of the deposition curve shown in Figure 10.1. We can also now appreciate why hygroscopic growth of particles shifts their pattern of deposition in the airways: as the particles grow they deposit, increasingly, in the larger airways.

10.6   Particle Metrics

We do not propose to discuss in any detail the many methods used to measure the concentrations of particles in the air. However, confusion may arise from the terms used to describe particle concentrations and we shall say a little about them.

If we imagine a machine which draws in air and sends it through some sort of filter which (1) collects particles and (2) can be removed to allow the mass that has been collected to be weighed, we will have imagined a standard method for measuring the mass concentration of particles in ambient air. Let us add a refinement. The collection system (the sampling head) might be modified to allow only particles of less than a specified size to be admitted.

In practice, modification of the sampling head is most important. The sampling characteristics of the head may be adjusted so as to mimic the sampling characteristics of the human respiratory system. We have seen that particles of more than 10 μm aerodynamic diameter do not, on the whole, penetrate beyond the upper airway (nasal passages, nasopharynx and larynx), particles of less than 10 μm aerodynamic diameter can enter the airways of the lung: from the trachea onwards into the branching system of airways that comprise the bronchi, bronchioles and alveoli. Particles that enter the airways of the lung are described as the "thoracic fraction" of the ambient aerosol and the mass concentration of such particles is monitored as PM_{10}. There is in fact a slight difference between the specifications of collection system designed to sample thoracic particles and PM_{10} but we shall not concern ourselves with this. PM_{10} means "the mass of particles of, in general, less than 10 μm aerodynamic diameter, per cubic metre". PM_{10} is therefore a metric of mass concentration. Of the particles monitored as PM_{10} the larger particles tend to deposit in the conducting airways (the airways that conduct air to the alveoli of the gas exchange zone of the lung); the smaller particles have a greater probability of reaching the gas exchange zone. In both cases some particles will be deposited.

Some, not all: see Figure 10.1! Measurements of mass concentration (PM_{10} for example) tell us about the mass concentration of particles in ambient air, they do not tell us the mass deposited in the lung. If one wishes to collect the particles most likely to be deposited in the gas exchange zone then a smaller size fraction should be chosen, smaller than that used for PM_{10} measurements. Deposition in the gas exchange zone peaks for particles of about 4 μm aerodynamic diameter. Such particles may be monitored by placing a sampling head with a "cut" at 4 μm. This means, in principle that all particles of <4 μm aerodynamic diameter are collected and all particles of less than 4 μm diameter are rejected. Unfortunately, such perfect selectivity is impossible to achieve.

$PM_{2.5}$ is also widely used as a metric of the ambient aerosol. Sampling heads with an acceptance curve characterised by 50% acceptance at precisely 2.5 μm aerodynamic diameter were developed to:

1. Give a reasonable representation of particles capable of reaching the gas exchange zone in normal adults.
2. Separate particles by source and physico-chemical characteristics.
3. Better reflect particles likely to deposit in the gas exchange zone in subjects with narrower than usual conducting airways: narrower, that is, than found in the normal adult. Narrow conducting airways mean more deposition of the smaller fraction of PM_{10} in these airways. Thus, though $PM_{4.0}$ might best characterise potential deposition in the gas exchange zone in normal adults, $PM_{2.5}$ might better characterise potential deposition in the gas exchange zone in children and in adults with diseases leading to narrowing of the conducting airways, for example asthma and chronic bronchitis. These groups are sometimes described as "vulnerable".

Figure 10.3 shows the acceptance curve of a sampler designed to allow measurement of $PM_{2.5}$.

It should be obvious that PM_{10} includes $PM_{2.5}$.

10.7 A Refinement of Interpretation of the Deposition Curve for Particles

Let us look again at the deposition curves shown in Figure 10.4.

Note, again, that the *y* axis shows the % deposition. This means the percentage of particles of the sizes specified on the *x* axis. In terms of

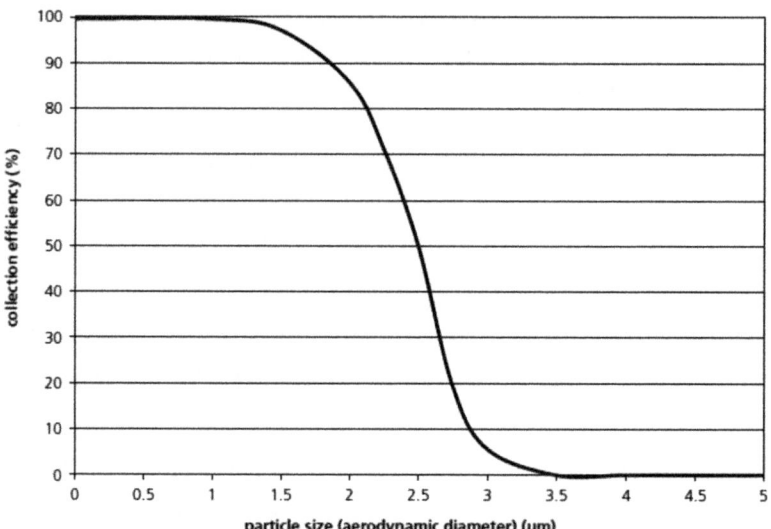

Figure 10.3 Acceptance curve of a sampler designed to allow measurement of PM$_{2.5}$.

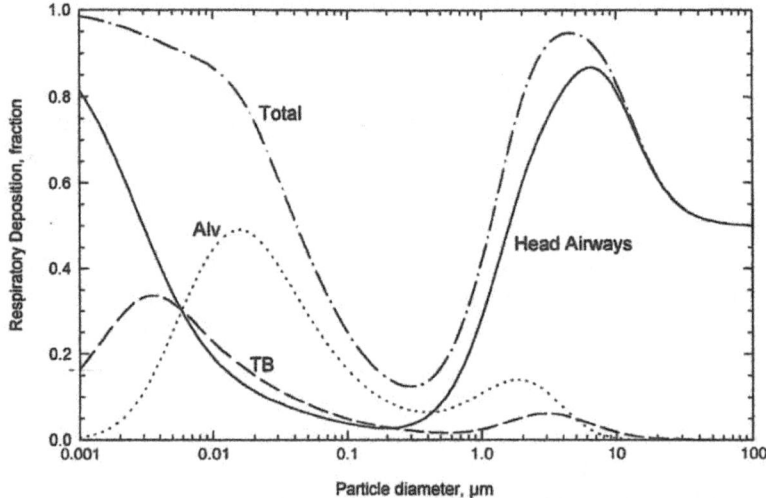

Figure 10.4 Respiratory deposition fraction against particle diameter.[1] Reproduced from ref. 1 with permission from John Wiley & Sons, Copyright © 1999, John Wiley and Sons Inc.

percentages the left hand hump of the pulmonary deposition curve looks much larger than the right hand hump. But in terms of deposited *mass* the right hand hump is more important than the left. Why is this? It is because the mass of a spherical particle is specified by the density of the particle and the cube of its radius. Let us compare the masses of a particle of 10 μm diameter with one of 0.1 μm diameter. The ratio of the diameters (or radii) is 100 to 1 but the ratio of the masses is 1 000 000 to 1. Or, a 10 μm particle has the mass of a million 0.1 μm particles.

We can now see that the great bulk of the mass deposited in the gas exchange zone is contributed by the right hand hump; the left hand hump contributes comparatively little mass. In other words, although very small particles deposit efficiently in the gas exchange zone they represent only a small fraction of the total mass deposited in that part of the lung. This is very important to the toxicologist. Let us assume that the particles being deposited in the gas exchange zone are soluble in water. We would expect them to dissolve in the thin layer of liquid that lines the gas exchange zone. If it is the mass that dissolves which drives the toxicological response then the right hand hump is more important than the left. Of course, if the particles are not soluble and if the very small particles can cross the gas exchange barrier into the blood stream and there, or on distant tissues, have some effects, they might be much more important than their deposited mass suggests.

References

1. W. Hinds, *Aerosol Technology*, John Wiley and Sons Inc, New York, 2nd edn, 1999.

11 Gases, Liquids and Droplets

11.1 Introduction

We have already said a good deal about particles; we now turn to gases, liquids and droplets. A number of major air pollutants, including ozone and nitrogen dioxide, occur as gases. Sulfur dioxide, for example, occurs as a gas and dissolves in water to produce sulfurous and sulfuric acids. Acid aerosols produced by sulfur dioxide dissolving in air-borne water droplets are less of a problem, today, in urban areas of the UK than they were in the coal-smoke era, but remain an important air pollutant in some parts of the world.

11.2 Gases

11.2.1 Critical Temperature and Pressure

All gases may be liquefied by the application of an appropriate pressure at an appropriate temperature. Each gas is characterized by a critical temperature; at temperatures greater than the critical temperature the gas cannot be liquefied by the application of pressure. Gases characterized by a critical temperature higher than room temperature are sometimes referred to as vapours. Each gas is also characterized by the pressure needed for liquefaction *at* the critical temperature: this pressure is defined as the critical pressure of the gas. Table 11.1 shows the critical temperatures and pressures of air and some common gases.

Basic Mathematics for Students of Air Pollutants
By Robert Maynard and Richard Atkinson
© Robert Maynard and Richard Atkinson 2024
Published by the Royal Society of Chemistry, www.rsc.org

Table 11.1 Critical temperatures and pressures of gases and air[a]

Gas	Critical temperature, T_c, °C	Critical pressure, P_c, (atm)
Air	−140.7	37.2
Nitrogen	−147.1	33.5
Oxygen	−118.8	49.7
Carbon dioxide	31.1	73.0
Carbon monoxide	−140.0	34.5
Nitrogen dioxide	158.4	100.0
Ozone	−12.1	54.6
Sulfur dioxide	157.8	77.7

[a]atm = atmosphere (in the sense of a unit of pressure). 1 atmosphere = 1013 mbar or 760 mm Hg or 100 kiloPascals (kPa).

11.2.2 Units Used to Describe the Concentration of Gases

Two systems are in general use: the mass per unit volume system, and the volume-fraction system. The former gives concentrations in the familiar mg m^{-3} or µg m^{-3} units; the latter gives concentrations as parts per million (ppm$_v$) or parts per billion (ppb$_v$). Note the subscript v as in ppb$_v$: this signifies that we are speaking of the volume fraction. It would be possible to use a volume-fraction *by mass*: ppm$_m$ and ppb$_m$. In the air pollution field, ppm and ppb, if unqualified, mean *by volume*.

Toxicologists prefer the volume-fraction system to the mass per unit volume system. This is because it is unlikely that the comparative toxicological activities of two gases will depend on the comparative masses of individual molecules but, rather, on their relative proportions in inspired air. Despite this, concentrations of gaseous air pollutants are often expressed as µg m^{-3}. The two systems are, of course, related and concentrations expressed in one system can be converted into concentrations expressed in the other.

In 1811 Avogadro put forward a hypothesis: equal volumes of all gases, under the same conditions of temperature and pressure, contain equal numbers of molecules. Hence, since 1 mole (or gram molecule) of any gas contains the same number of molecules it must occupy the same volume, at standard temperature and pressure (STP: 273 Kelvin and 760 mm Hg pressure). Let the density of a gas at STP be x g L^{-1}, let the gram molecular weight of the gas be y grams. Then the volume occupied by 1 gram molecule of gas will be y/x L. Study of a range of gases has shown that $y/x = 22.41$ L.

The gram molecular weight (GMW) of a compound, in this case a gas, is the molecular weight expressed in grams. Thus the GMW of, for example, ozone (O_3) is $3 \times 16 = 48$ g.

It follows that:

The volume occupied by 48 g of ozone is 22.41 L (at STP).

1 L is occupied by 48/22.41 g of ozone.

1 cm^3 is occupied by 48/22.41 mg of ozone.

Let us say that the concentration of ozone is 1 cm^3 m^{-3}.

The mass concentration of ozone is therefore 48/22.41 mg m^{-3}.

Now then, 1 cm^3 m^{-3} = 1 ppm (there are 1 million cubic centimetres in 1 cubic metre: think this through, 1 m = 100 cm, 1 m^3 = 100^3 cm^3 = 1 million cm^3).

So, 1 ppm$_v$ = 2.14 mg m^{-3}

and, 1 ppb$_v$ = 2.14 μg m^{-3}

or, 1 mg m^{-3} = 1/2.14 = 0.47 ppm$_v$

and, 1 μg m^{-3} = 1/2.14 = 0.47 ppb$_v$.

We have, of course, been working at STP. If we wish to work at other temperatures and pressures we shall have make a correction.

We defined the volume occupied by 1 gram molecule of a gas as 22.41 L at STP. This is the molar volume (MV) of the gas. MV is corrected to the volume (V) occupied at T degrees K and P mm Hg by:

$$V = 22.41 \times \frac{T}{273} \times \frac{760}{P}$$

If P is expressed in millibars then 1013 should be substituted for 760 in the above equation.

For liquids suspended as an aerosol in air it seems obvious to express concentration as mass per unit volume of air. But a mass per unit mass system is sometimes used. Let us think through this.

Air is 80% nitrogen and 20% oxygen (in round figures!).

Can we calculate an "equivalent GMW" for air?

The GMW of nitrogen is 28 g.

The GMW of oxygen is 32 g.

The "GMW" of the mixture will be (0.8 × 28) + (0.2 × 32) = 28.8 g.

We have written "GMW" because GMW applies to specific molecules.

We can now say that 28 g of air occupies 22.41 L.

Or, that 1 L of air weighs 28/22.41 g.

Or, that 1 m^3 of air weighs 28/22.41 × 1000 g = 1.25 kg.

If the concentration of some liquid suspended in air is given as 1 μg m^{-3} we can say that the concentration is 1 μg per 1.25 kg.

That is, 1/1.25 μg kg^{-1}.

Recalling that 1 kg = 1 × 10^9 μg,

we may say that the concentration is 1/1.25 μg/1 × 10^9 μg.

Recalling that 1 billion = 1×10^9 (the "English billion", 1×10^{12} is not used in scientific work), then, 1/1.25 ppb by mass, that is 1 ppb$_m$ or,

1 ppb$_m$ = 1.25 µg m^{-3}.

Of course we must remember, again, that the figure 22.41 L applies to STP and that we should have to correct this if we are working at ambient temperature and pressure.

11.2.3 The Diffusion of Gases

The rate at which a gas diffuses is defined by Fick's first law of diffusion which defines the diffusion coefficient, D, as the quantity of substance diffusing *per unit time* through *unit area* of, for example, a barrier across which diffusion is taking place, under a *unit concentration gradient*. If we compare two gases diffusing across the same barrier then the gas with the lower molecular weight will diffuse more rapidly than the gas with the greater molecular weight. This is an expression of Graham's law of diffusion which states that if D_1 is the rate of diffusion of gas 1, which has a molecular weight of MW$_1$, it is related to the rate of diffusion of gas 2 (D_2), which has a molecular weight of MW$_2$, by

$$\frac{D_1}{D_2} = \frac{\sqrt{MW_2}}{\sqrt{MW_1}}.$$

We also need to know about Henry's law of solubility of gases. This states that the concentration of a gas in solution in a liquid is equal to the partial pressure of the gas in the solution multiplied by the solubility of the gas in the solution.

We have mentioned the term partial pressure, let us define it. Dalton's law of partial pressures states that in a mixture of gases exerting a pressure P each gas will exert a pressure equal to a fraction of P and that fraction will be proportional to the fractional concentration of each gas in the mixture.

Thus, if we consider dry air at STP: the pressure = 760 mm Hg.

The mixture is, in round figures, 79% nitrogen and 21% oxygen.

Thus, the partial pressure of nitrogen will be $0.79 \times 760 = 600$ mm Hg and the partial pressure of oxygen will be $0.21 \times 760 = 160$ mm Hg.

Consider a liquid containing no gas placed in contact with a gas. Molecules of gas will diffuse into the liquid until gas molecules leave and enter the liquid at the same rate: until the gas and liquid phases are in equilibrium. The partial pressures of the gas in the gas phase and liquid phase will be identical. Now then, consider a system where two different liquids are exposed to the same partial pressure of gas. At equilibrium the partial pressures of the gases in the liquids will be identical and equal to the partial pressure of the gas in the gas phase. But the concentrations of the gases may well differ between the two liquids. Rather obviously the concentration in the liquid in which the gas is very soluble will be higher than the concentration in the liquid in which the gas is rather insoluble. The fact that the partial pressures are identical means that gas molecules are just as likely to escape from the first liquid as they are from the second liquid. This capacity to escape, this escaping tendency, reflects what physical chemists call the "activity" of the gas molecules. It can also be described as the fugacity of the gas molecules: fugacity has the same etymological root as fugitive meaning a tendency to flee. Activity is defined by partial pressure and not by concentration.

Now we are ready to answer a tricky question. Do molecules of gas always diffuse down their concentration gradient? The obvious answer is, yes. But this is the correct answer only when other factors are constant. Sometimes a gas can diffuse up its concentration gradient. At first sight this is difficult to believe. Let us consider a hypothetical system. Let us imagine four spaces numbered 1 to 4 as shown in Scheme 11.1. Let us assume that the barrier between the liquid phases (spaces 3 and 4) and between the gas phases (spaces 1 and 2) are impermeable to the gas. The barriers between the gas and liquid phases are assumed to be permeable to the gas.

P stands for partial pressure, x stands for a gas, S stands for solubility, C stands for concentration.

Now then if we changed the barrier between space 3 and space 4 so that the gas, x, could diffuse across the barrier which way would the gas diffuse? From space 3 to space 4 or from space 4 to space 3?

Space 1 Gas phase $Px = 100$ mm Hg	Space 2 Gas phase $Px = 10$ mm Hg
Space 3 Liquid phase: liquid A $Px = 100$ mm Hg $Sx = 1$ mg m^{-3} per 1 mm Hg partial pressure $Cx = 1 \times 100 = 100$ mg m^{-3}	Space 4 Liquid phase: liquid B $Px = 10$ mm Hg $Sx = 100$ mg m^{-3} per 1 mm Hg partial pressure $Cx = 100 \times 10 = 1000$ mg m^{-3}

Scheme 11.1 Hypothetical system comprising four spaces, two liquid and two gas.

The answer is that the gas, *x*, would diffuse from space 3 to space 4: down its partial pressure gradient and up its concentration gradient.

Surprised? Most people are when they first meet this fact. What has foxed us is that we are dealing with two different liquids: the gas is much more soluble in liquid B than in liquid A. Is this important? Well, it is to fresh-water fish and to physiologists who study fish: gases diffuse across the gill epithelium from blood (containing a good deal of salt) to fresh water. The solubility of oxygen is greater in fresh water than in fish blood.

We can now face a second tricky question. We all know that, in the lung, oxygen diffuses from the smallest air spaces, the alveoli, to the blood and that carbon dioxide diffuses in the opposite direction. The molecular weights (32 and 28 respectively) are not sufficiently different to make a significant difference as regards the diffusion coefficients: we can ignore the effect of Graham's law. What we cannot ignore is that carbon dioxide is 24 times as soluble as oxygen in body fluids. In terms of how quickly the individual molecules of oxygen and carbon dioxide diffuse there is nothing in it. Why, then, do physiologists say that carbon dioxide diffuses 24 times as rapidly as oxygen?

The answer is that physiologists use a special "coefficient of diffusion". This is the Krogh coefficient of diffusion or, better, Krogh's constant of diffusion. August Krogh (1874–1949) was one of the greatest of respiratory physiologists. He defined his constant as: the amount of gas diffusing per *unit time*, across *unit area* of the barrier, per *unit (mm Hg) difference in partial pressure*. Ah! *Per mm Hg difference in partial pressure, NOT per unit difference in concentration.* We now have the answer. Carbon dioxide is 24 times as soluble as oxygen and thus for a 1 mm Hg difference in partial pressure there will be a 24 fold difference in concentration. So, the molecules of oxygen and carbon dioxide are moving at about the same rate but there are 24 times more molecules of carbon dioxide than molecules of oxygen on the move for a unit difference in partial pressure. We can now see why physiologists say carbon dioxide diffuses 24 times as rapidly as oxygen.

11.3 Liquids

Everybody knows that liquids evaporate. If a volume of liquid is placed in an enclosed space evaporation will take place: molecules leave the surface of the liquid phase. Molecules will also leave the gas phase and re-enter the liquid phase. Equilibrium is reached when molecules leave and re-enter the liquid phase at the same rate. In the equilibrium state, the air above the liquid is said to be saturated with vapour and the vapour exerts a pressure equal to the saturated vapour pressure (SVP) of the liquid. SVP depends on temperature which controls the rate at which molecules move.

The SVP of a liquid defines its volatility: the higher the SVP then the greater the volatility of the liquid. When SVP is equal to atmospheric pressure the liquid boils. The SVP of water at 100 °C is 760 mm Hg: water boils at 100 °C at sea level. A moment's thought will suggest that water will boil at less than 100 °C as the atmospheric pressure falls below 760 mm Hg. It is impossible to make tea at the top of a high mountain: the water boils at too low a temperature to brew the tea. Of course, air is not always saturated with water vapour. When saturation is complete the pressure exerted by the vapour is the SVP. The relative humidity (RH) of air defines the extent of saturation. Thus:

$$RH = actual\ vapour\ pressure / SVP \times 100$$

We have noted that SVP is dependent on temperature; it is important to remember that SVP is independent of barometric pressure. Of course, the liquid and gas phases must be in contact for these rules to apply.

If a quantity of water is introduced into a rigid evacuated container at 37 °C evaporation will occur and the pressure in the container will rise to 47 mm Hg: the SVP of water at 37 °C is 47 mm Hg.

If a quantity of water is introduced into a rigid container of dry air at 760 mm Hg at 37 °C the water will evaporate and the pressure in the container will rise to 760 + 47 = 807 mm Hg.

If, on the other hand, the contents of the container were maintained at atmospheric pressure (760 mm Hg) by allowing the container to expand, then the water vapour would still come to exert a pressure of 47 mm Hg, the other gases exerting a pressure of 760 − 47 = 713 mm Hg (changes in volume and pressure are assumed to occur isothermally). The gases other than water vapour would obey Dalton's law of partial pressures and exert pressures (the partial pressures of the individual gases) in accordance with the volume proportion occupied by each gas. Thus:

Assuming that air is 21% oxygen, the partial pressure of oxygen would be 0.21 × 713 = 149.7 mm Hg.

This is an important fact in respiratory physiology: the partial pressure of oxygen (PO_2) in dry air is 159.1 mm Hg but the PO_2 in inspired air, which is saturated with water vapour in the upper airways of the lung, is 149.7 mm Hg.

The maximum concentration of a substance in the vapour phase (saturated vapour concentration, SVC) may be calculated from the SVP. This can be done, to a useful level of accuracy, by application of the ideal gas equation:

$$PV = nRT$$

Let us define our terms – Table 11.2.

From the ideal gas equation:

$$n = PV / RT$$

Let us consider water at 37 °C (310 degrees K), SVP = 47 mm Hg = 47 × 101 325/760 = 6266.2 N m^{-2} (the reader may have wondered about the conversion of mm Hg to N m^{-2}. 760 mm Hg = 100 kPa = 101 325 N m^{-2}).

Then, considering 1 m^3 of air saturated with water vapour at 37 °C (310 degrees K)

$n = (6266 \times 1)/(8.3143 \times 310) = 6266/2577.4 = 2.43$ mole m^{-3}

1 mole (gram molecule) of water = 18 g.

Thus the concentration of water in air saturated with water vapour at 37 °C is

$2.43 \times 18 = 43.8$ g m^{-3}.

The numerical similarity between SVP and SVC, in this calculation, is due to the fact that

101 325/760 × 1/8.3143 × 1/310 × 18 = 0.93.

We should not expect such numerical similarity for substances with other molecular weights.

We could attack the question in another way.

Let us recall that 1 mole of any gas, at STP, occupies 22.41 L.

1 mole of water vapour, at STP, occupies 22.41 L.

1 mole of water vapour, at 37 °C and 760 mm Hg occupies 22.41 × 310/273 = 25.4 L.

Table 11.2 Definition of terms[a]

Symbol	Term	Units
P	Pressure	N m^{-2}
V	Volume	m^3
n	Number of moles of gas per m^3	—
T	Absolute temperature	degrees K
R	Gas constant: 8.3143	N m K^{-1} mol^{-1}

[a]N = the Newton; N m^{-2} is a unit of pressure: one Newton per square metre.

Table 11.3 Saturated vapour pressure and concentration of water at a range of temperatures.

Temperature/°C	SVP/mm Hg	SVC/g m^{-3}
0	4.57	4.87
10	9.14	9.36
20	17.36	17.15
30	31.51	30.08
40	54.87	50.67

This means that 1 mole of water (18 g) if constrained to occupy 25.4 L would exert a pressure of 760 mm Hg.

But, we know that the SVP of water at 37 °C is only 47 mm Hg.

If we took one mole of water vapour and allowed it to expand its pressure would fall. What would the volume need to be for the pressure to be 47 mm Hg?

It would need to be 760/47 × 25.4 = 410.7 L.

This means that 410.7 L of air saturated with water at 37 °C would contain 18 g of water.

Or 1 L of air saturated with water would, at 37 °C, contain 18/410.7 g of water.

Or 1 m^3 of air saturated with water would, at 37 °C, contain 18/410.7 × 1000 g of water.

That is 43.82 g m^{-3}, the same answer as we found above. Table 11.3 gives the SVP and SVC for water at a range of temperatures.

11.4   Fog

If the temperature of a mass of air containing water vapour falls below the temperature of saturation water will condense onto the surfaces of dust particles and fog will form. If air saturated with water vapour at 30 °C is cooled to 10 °C then, as we can see from Table 3.4, some 20 grams of water will be condensed from each cubic metre of air. Under suitable conditions this water forms fog. If the layer of air near the ground cools, for example, during the night, then water will condense onto the surface of the ground and plants and dew is formed. If the dew freezes then frost is produced. The dew point is defined as the temperature of 100% saturation of the air with water vapour. For example, air 60% saturated with water at 40 °C will contain about 33 g of water per cubic metre. This content is equivalent to 100% saturation at 32 °C. If the air is cooled from 40 °C, the level of saturation will rise to 100% at 32 °C. Further cooling will lead to condensation: the dew point of air which is 60% saturated with water at 40 °C is 32 °C.

11.4.1 The Fate of Liquid Aerosols

It may be shown that the vapour pressure exerted by a liquid at a convex surface, for example at the surface of a droplet, is greater than that exerted at a plane surface. This is the Kelvin effect and is very important in considering the stability of liquid droplets. If a very small droplet occurs in an atmosphere in which the vapour pressure (of the liquid forming the droplet of course) is at the maximum associated with a plane surface (that is SVP) then the droplet will shrink. Why?

The vapour pressure at the surface of the droplet is greater than SVP and liquid will evaporate from the surface even though the surrounding air is saturated with water. Loss of liquid will cause the droplet to grow smaller. As it does so the convexity of its surface will increase and the vapour pressure at the surface will rise. More liquid will evaporate and the droplet will grow ever smaller. Of course this would also happen if the air surrounding the droplet was of a relative humidity of less than 100%.

Now then, let us consider the more interesting but more complicated problem of a droplet of a watery solution of some non-volatile substance such as sodium chloride. As water evaporates from the surface of such a fluid the concentration of the solute will, inevitably, increase. Now then, vapour pressure is a colligative property of solutions (along with boiling point, freezing point and osmotic pressure it depends on the number of molecules present per unit volume of solution) and as the concentration of solute rises the vapour pressure at the surface will fall. The combination of vapour pressure tending to increase because the surface is becoming more convex (as the droplet becomes smaller) and the vapour pressure tending to fall because the concentration of solute is increasing means that a point of stability will be reached. The same argument, of course applies in reverse, to a droplet growing larger. We can now imagine that at a certain solute concentration and relative humidity a droplet would become stable; that is in equilibrium with its surroundings. Cocks and Fernando[1] calculated that droplets of a 20% solution of sulfuric acid would be in equilibrium at a relative humidity of 88% at a temperature of 37 °C.

Cocks and Fernando also undertook a computer simulation of the growth of droplets of on passing from ambient air to the water-saturated air of the respiratory tract. Droplets of 0.1 and 1.0 µm radius and containing 20%, 40% and 60% solutions of sulfuric acid were considered. The particles grew and the concentrations of sulfuric acid within the droplets fell. Growth of the droplets would be expected to change their pattern of deposition in the respiratory tract: see Chapter

10. An increase in size increases the likelihood of deposition by sedimentation in the larger airways of the lung and decreases the likelihood of deposition by diffusion in the smaller airways and air spaces of the lung. In addition to the reduction in the concentration of sulfuric acid resulting from the uptake of water by the droplets, neutralization of acid by absorption of ammonia, produced in the mouth by bacteria would also be likely to occur. This was modelled by Cocks and McElroy.[2] The original papers by these authors should be consulted for details of their methods and their findings. The authors concluded that conditions in the then current (1984) urban setting differed very significantly from those experienced in the London fogs in earlier times. They argued that the then current (1984) concentrations of sulfuric acid rarely exceeded 50 µg m^{-3} "even in heavily polluted areas such as Los Angeles". Their model predicted that even with nasal breathing (the amount of ammonia produced in the nose is less than that produced in the mouth) substantial neutralization of the inspired acid would occur. They continued, "Thus if acidity were the predominant factor injurious to health, little effect would be expected." They referred to work by Kerr et al.[3] which showed that a concentration of sulfuric acid of 100 µg m^{-3} in the (droplet) size range of 0.1–0.3 µm in ambient air has no adverse effects after exposure for 4 h. They noted that in the London fogs (smogs) of earlier times sulfuric acid concentrations of 1000 µg m^{-3} had been recorded and that droplet sizes had been relatively large and thus slow to neutralize. The bottom line, so to speak, of their argument was that it would be unwise to extrapolate from findings of effects on health during the London fogs to present day conditions.

References

1. A. T. Cocks and R. P. Fernando, The growth of sulphate aerosols in the human airways, *J. Aerosol Sci.*, 1982, **13**(1), 9–19.
2. A. T. Cocks and W. J. McElroy, Modelling studies of the concurrent growth and neutralization of sulfuric acid aerosols under conditions in the human airways, *Environ. Res.*, 1984, **35**, 79–96.
3. H. D. Kerr, T. J. Farrell and L. R. Sauder, *et al.*, Effects of sulfuric acid aerosols on pulmonary function in human subjects: an environmental chamber study, *Environ. Res.*, 1981, **26**, 42–50.

12 Elementary Considerations of Line Fitting Techniques: Derivation of Concentration-Response Relationships

12.1 Introduction

This chapter explains some elementary concepts about the fitting of lines to data obtained by experiment or from observational epidemiological studies. The purpose of "fitting lines" is to describe, in terms of a mathematical equation, the relationship between two (or more) continuous variables. The more formal term for this process is "regression".

12.2 Preliminary Considerations

Our first task in understanding the relationship between variables is to decide which variable is the *dependent* variable, *i.e.* the variable whose values depend upon the values of other, *independent*, variable(s). These decisions are determined from knowledge of the underlying science/processes involved. For example, "land use" regression models are used to estimate pollution concentrations based upon (amongst other factors) traffic density. Pollution concentrations are dependent upon traffic emissions (which are, in general, independent of pollution levels).

Basic Mathematics for Students of Air Pollutants
By Robert Maynard and Richard Atkinson
© Robert Maynard and Richard Atkinson 2024
Published by the Royal Society of Chemistry, www.rsc.org

Our next task is to plot the data using a scatter plot. A scatter plot is a graph showing each data point, the location of which on the plot is determined by the values of each variable. Let us look at an example.

Figure 12.1a shows some hypothetical data: x is the independent variable (usually plotted on the horizontal x-axis or abscissa), y, the dependant variable (usually plotted on the vertical y-axis or ordinate) which varies with values of x. It is obvious that the data suggest that a straight line describes the variations of y with variations of x.

If we took a rule and drew in a straight line through the points, by eye, we would have "fitted" a straight line to the data, see Figure 12.1b. This means that we have decided that a straight line is the appropriate *model* for the relationship between the two variables. We would have also determined, by eye, the slope, or gradient, of the line and the intercept (the value of y when $x = 0$; assuming of course that the x values go down to 0).

Now, we could have joined the points and produced a zig-zag relationship. Why did we reject this and choose a straight line? The usual reason is that we could not explain the zig-zag relationship, we think it unlikely to be a true representation of the underlying

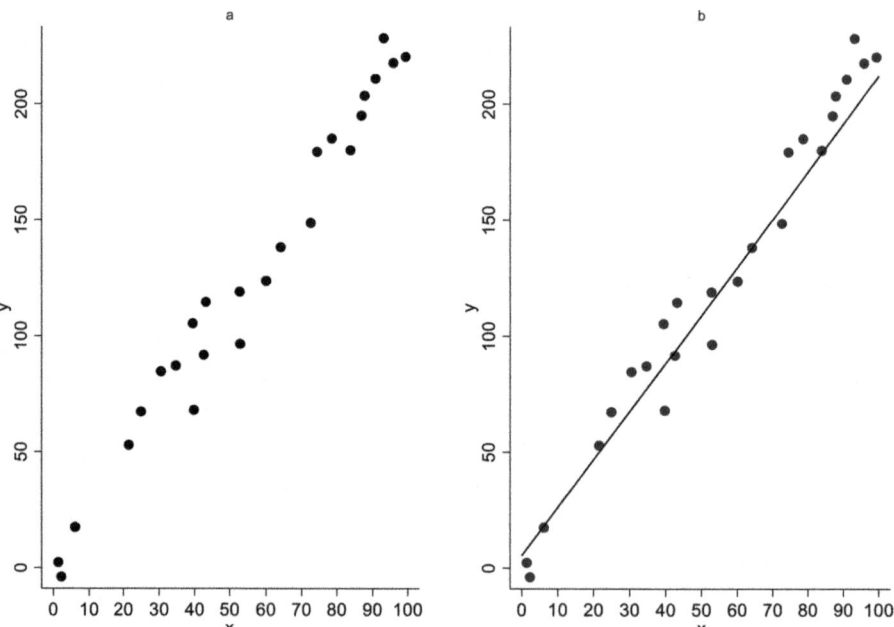

Figure 12.1 Scatter plot of hypothetical data showing the relationship between two variables, x and y, figure (a). Figure (b) shows a straight line that has been fitted to the data.

(true) relationship between the two variables and prefer to think that errors, perhaps errors of measurement, have caused the distribution of points around the straight line. Most readers will know that the technique of linear regression replaces the use of a rule and allows us to determine the equation of the straight line which best fits the data. The key word here is "best": many lines might fit the data; we are looking for the line which provides the closest or best fit. It is important to remember that all models are wrong; however, some are useful.

12.3 Linear Regression

In mathematical language the equation for the linear model is:

$$Y = \alpha + \beta X + \varepsilon$$

where α is the intercept on the y axis; β the slope, or gradient, of the line; and ε an error term. It is the error term that allows for and describes deviations of real values from the theoretical values described by the linear relationship. The process of determining or estimating α and β from the data involves selecting values (of α and β) that minimise the deviations of the observed data from the fitted line. This process makes a number of key assumptions about the errors: (i) they are independent of each other; (ii) they are normally distributed; and (iii) they are independent of X (homoscedastic). We will not go into the details of this process; instead, we refer the reader to statistics textbooks that explain how linear regression works: see Further Reading. Suffice to say that many computer software packages will fit the regression line for us.

Assuming we have carried out a linear regression calculation we now have the equation (characterised by the slope "b", and intercept "a") of the straight line that best fits the data, meaning that the line is drawn so as to minimise the deviations of the points from the fitted line. We have improved on the use of a rule in the sense that we can now justify the equation of the line, but we have remained committed to the assumption that the data can be described by a straight line.

It is worth noting that "a" and "b" are specific to the data under consideration. If we obtained another sample of data and repeated the regression analysis we would obtain new values, let's say "a_1" and "b_1". Both "a" and "a_1" are estimates of the hypothetical parameter α; likewise, "b" and "b_1" are estimates of β. As with other parameter estimates (such as the mean we met in Chapter 5), we can calculate confidence intervals (CI) for them and we interpret them as before

(Chapter 6). A confidence interval for "*b*" that includes 0 is of special interest. Why? Because if the CI contains 0 then the true relationship between *Y* and *X* could be a horizontal line implying that the value of *Y* is not dependent on *X*.

How does the fitted line help us? First, it describes the relationship between the two variables. How quickly does *y* change for a unit change in *x* (the slope) and what is the value of *y* when *x* is zero (the intercept). The estimated equation also enables us to predict, within the range of the data, the (average) value of *y* for any value of *x* (not just for the actual values of *x* in the dataset). In the previous sentence we included the qualification "within the range of the data". This is an important condition and worthy of further consideration. It means that the straight line we have fitted to the data holds only for the range of *x* values represented in our data. We have no information on how the two variables are related outside of this range. We must therefore appreciate and accept this limitation and not extrapolate the relationship beyond the range of the data.

Linear regression can easily be extended to include multiple independent variables. The equation extends naturally as follows:

$$Y = \alpha + \beta_1 X_1 + \beta_2 X_2 \ldots + \beta_p X_p + \varepsilon$$

where *p* is the number of independent variables. We can also incorporate categorical variables and multiple interaction terms on the right-hand side of the equation. We will not pursue these topics further but instead refer the interested reader to a selection of excellent texts on the topic: see Further Reading.

12.4   Correlation Coefficient

At this point it is appropriate to mention a related concept: the "correlation coefficient"; more formally the "Pearson" or "Pearson product-moment" correlation coefficient. One often hears or reads that two variables are correlated. Such a statement may be qualified further by use of adjectives such as weakly/strongly and negatively/positively. What do these terms mean? The correlation coefficient, usually denoted by the letter "*r*", is a measure of how well, or how strongly, the data conform to a straight line. "*r*" takes a value from -1 to 1; if $r > 0$ then *y* is said to be positively correlated with *x*; *y* increases as *x* increases. If $r < 0$ (that is, *r* has a negative value) then *y* is said to be negatively correlated with *x*; *y* decreases as *x* increases. The size

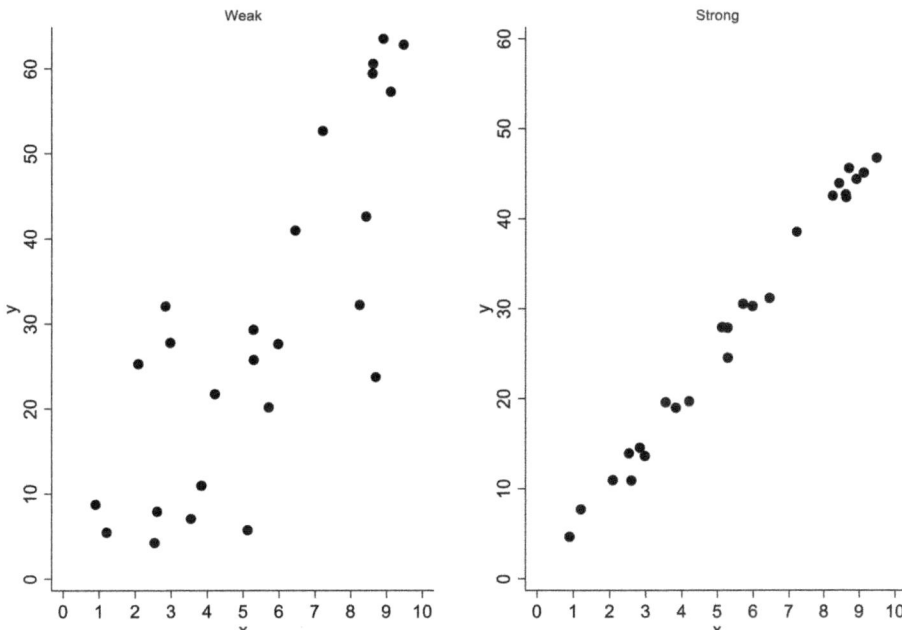

Figure 12.2 Examples of weak and strong positive correlation between two variables.

of r indicates the strength of the linear relationship. A weak (positive or negative) correlation is indicated by small (absolute) values of r; a strong (positive or negative) correlation is indicated by large (absolute) values. Figures 12.2 and 12.3 show examples of weak and strong, positive and negative correlation respectively. Note, r tells us nothing about the magnitude of the slope of the regression line.

Finally, if $r = 0$, then y and x are not correlated and there is no linear relationship between the two variables (Figure 12.4).

12.5   Sums of Squares

We have mentioned "theoretical considerations" more than once, what do we mean? We may have a theory to explain the relationship between x and y: that theory might suggest a straight line or, on the other hand it might suggest a logarithmic or some other non-linear relationship. By plotting the data as described above we are testing our theory. If the plot fails to produce the result we expect we may need to rethink the theory. We should distinguish between theories based on what we regard as fundamental laws of physics from theories, of our

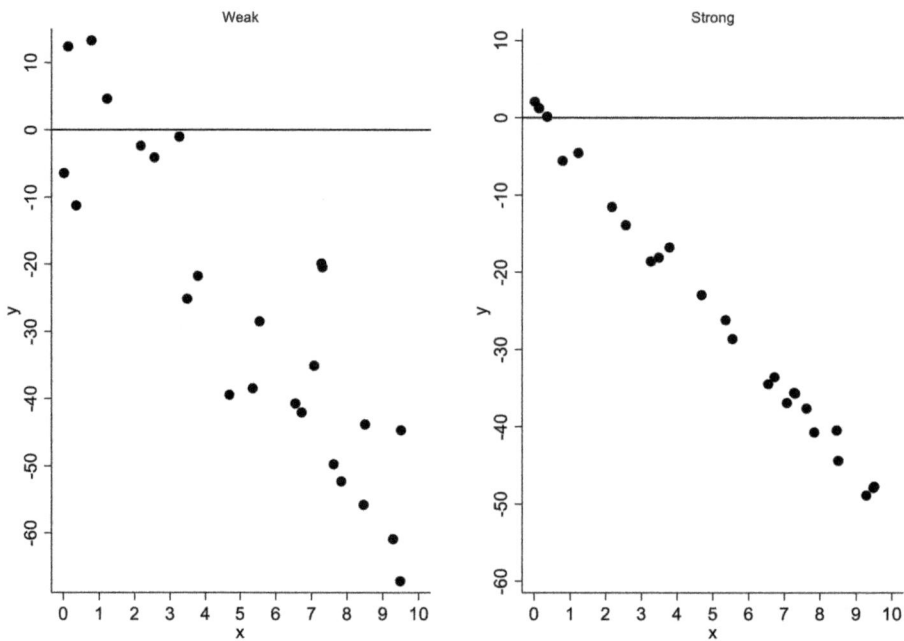

Figure 12.3 Examples of weak and strong negative correlation between two variables.

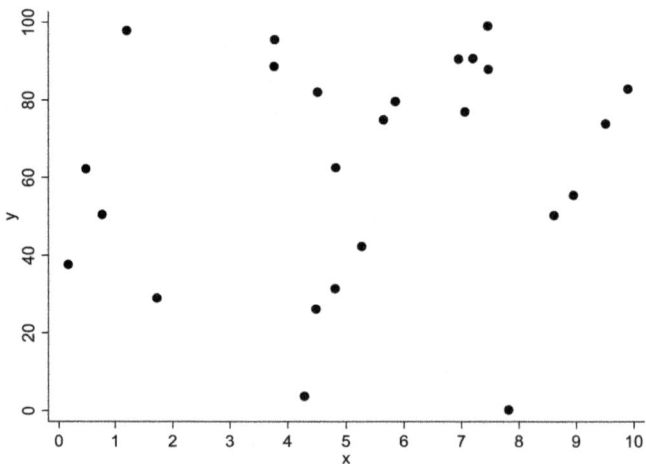

Figure 12.4 Examples of two variables showing very little correlation.

own invention, that we have put forward to explain the relationship between the variables. In the former case if the data do not fit a prescribed line we might think there was something wrong with the data; in the latter case we might think there was something wrong with the explanation or theory.

We might now consider a different case: let us assume we have no preconceived ideas about the relationship between x and y. When we plot the data we may, as above, immediately see that the data fits some mathematical relationship that can be easily recognised and for which a standard equation is available. On the other hand, it may not. Now we have a problem. Should we assume a specific mathematical relationship or simply let the data define the line? If we decide to assume a specific form of mathematical relationship then we might reasonably be described as forcing the data to the shape of the preconceived relationship. We might wonder whether we are justified in doing so.

One way of answering this question is to look at the r^2 value. Let's think this through. Imagine a series of values of x and a corresponding series of values of y. After calculating the equation for the line linking the xs and the ys, the linear regression equation, for each value of x we will have an estimated value of y, let's call that value y_e. We now ask: how much of the variation of y around the mean of all the values of y can be explained by the variation in y defined by the regression line. Let us accept the following equations, they can be found in any textbook of statistics.

The dispersion (another word for variation) of the estimated values of y_e around the mean of all the values of $\bar{y}$ is given by

$$\Sigma\left(y_e - \bar{y}\right)^2$$

Statisticians call this SSR meaning the sum of the squares for regression.

The dispersion of the actual values of y around the estimated values of y (y_e) is given by

$$\Sigma\left(y - y_e\right)^2$$

Statisticians call this SSE meaning the sum of the squares for error.

The dispersion of the actual values of y about the mean of all the values of y is given by

$$\Sigma\left(y - \bar{y}\right)^2$$

Statisticians call this SS meaning the total sum of the squares.

SS = SSR + SSE (for proof of this relationship see Further Reading: Alder and Roessler).

The square of the correlation coefficient, also called the coefficient of determination, is given by

$$r^2 = \frac{\text{SSR}}{\text{SS}}$$

What does this mean?

It means that r^2 tells us what fraction of the total dispersion of y about its mean can be explained by the dispersion of the predicted values of y about the mean of all the values of y. We could express r^2 as a percentage by multiplying the above formula for r^2 by 100. The higher the values of r^2 the more of the dispersion of y about the mean of all the values of y can be explained by the regression equation. If all the points lay on a straight line then r^2 would equal 1, or 100%. If the value of r^2 were 0.1, or 10%, we would have serious doubts about the explanatory power of the regression line.

12.6   Non-linear Regression

Let us imagine now that inspection of the data presented in the scatter plot suggests that the relationship between the dependent and independent variables do not conform to a straight line. Figure 12.5a suggests that the data conform to a curve. Daily average temperature over 12 months is an example of data that exhibit such a pattern. The curve will be familiar to many readers, it is a sine curve. If we decide that it makes good sense, on theoretical grounds that the data follow a sine curve then statistical techniques are available which allows the sine curve that best fits the data to be drawn (Figure 12.5b).

Figure 12.6 suggests another sort of curve. We have met this curve already (Chapter 3): it represents an exponential relationship. This means that if we plotted the logarithm of the values on the y axis against the values on the x axis, plotted on an ordinary arithmetic scale, then we would obtain a straight line. Indeed, we would if the data really did conform to an exponential relationship. Of course, it might be that the fit of the data to an exponential relationship is not perfect and the desired straight line will not be quite straight. But for practical purposes let us assume the log-plot produces a straight line. What have we achieved? In technical terms we have used a transform: we have transformed the data by plotting the logarithms of the values on the y axis. Other transforms exist. We should recall that we decided that the data looked as though a log-plot would produce a straight line:

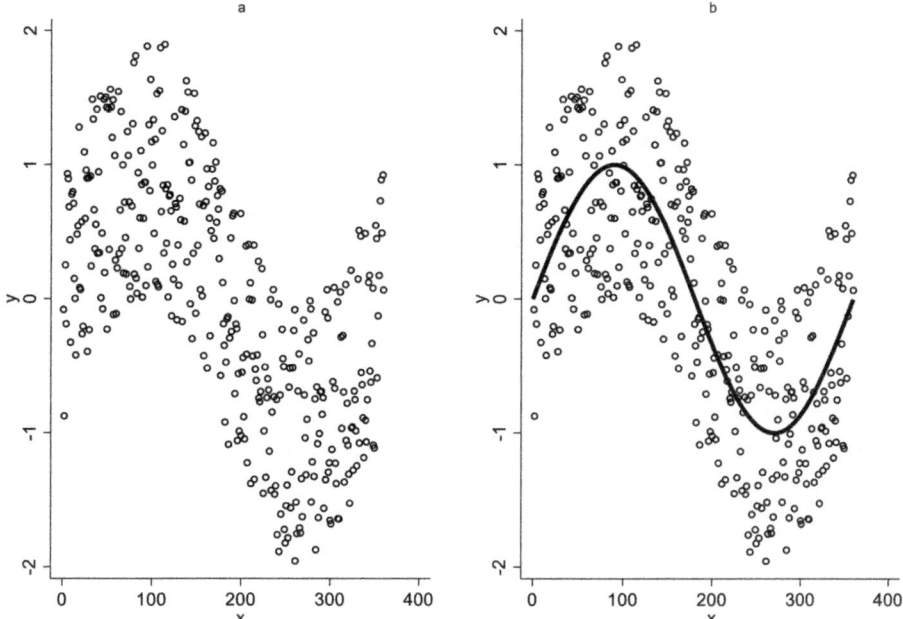

Figure 12.5 Scatter plot of hypothetical data exhibiting a sinusoidal relationship between two variables, *x* and *y*, figure (a). In figure (b) a sinusoidal curve has been fitted to the data.

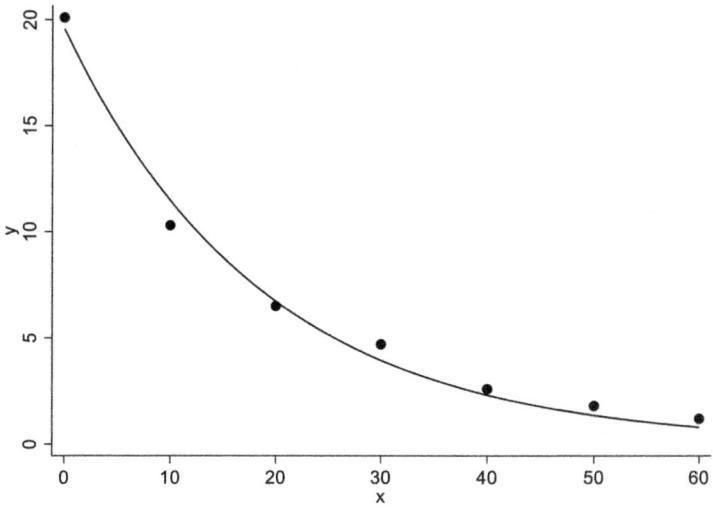

Figure 12.6 Example of an exponential relationship between two variables.

this decision was not based on theoretical considerations, merely on the appearance of the data.

12.7   Other Forms of Regression

So far in this section we have focused on linear regression – the dependent variable as a linear function of the independent variable(s). There are other types of regression analysis. In the air pollution field today, logistic and Poisson regression are important techniques. Both are examples of generalised linear regression. This advanced regression technique allows us to model alternative dependent outcomes such odds ratios (logistic) and counts of events (Poisson), each assuming specific error distributions. We do not propose to go into these techniques in detail but do provide below an outline so that the reader can appreciate some of the key features of these forms of regression.

12.8   Logistic Regression

Air pollution has been associated with a wide range of diseases and health outcomes. Some outcomes are based not upon a continuous measurement but upon categorisation. For example, when studying asthma, we might categorise individuals as either asthmatic or non-asthmatic. The outcome is binary. For the purpose of statistical analysis, we can represent these outcomes using a variable taking the values 1 or 0. In our example we would therefore assign the value 1 to asthmatic individuals and 0 to non-asthmatics. We are then interested in the probability that an individual is coded 1 and, furthermore, what individual characteristics might influence that probability. Logistic regression is a statistical tool that allows us to model the relationship between categorical outcomes and explanatory variables and is widely used. Pharmacologists use this approach when considering the interaction between drugs and drug receptors, an area of pharmacology called pharmacokinetics. Let us take a short break from air pollution science and see how pharmacologists think about such problems. The reader who is not interested in pharmacology may skip the algebra and read on from where we pick up the argument with the formula

$$\log\left[\frac{r}{1-r}\right]$$

We shall not go into the pharmacology of receptors in any detail; let us simply accept that receptors, we should say drug receptors, are molecules with which drugs can interact. Interaction leads to some change in function of the cells bearing the receptors and thus to the pharmacological activity of the drug. The interaction between a drug and the receptor with which it interacts can be analyzed as follows.

Let a drug (an administered chemical) be represented by D.

Let the receptor be represented by R.

Then the reaction between the drug and the receptor can be represented by

$$D + R \rightleftharpoons DR$$

This means that D and R react with each other the produce DR but, also, that DR breaks down to reform D and R. We should make clear that DR stands for the combination of drug and receptor and not D multiplied by R.

The law of mass action states that the rate of the forward reaction (the formation of DR) can be defined by $[D][R] \times k_1$.

k_1 is a constant, the rate constant, of the reaction. $[D][R]$ means the concentration of the drug, D, multiplied by the concentration of the receptor, R. Readers will know that chemists use square brackets to indicate concentration.

Similarly, the reverse reaction (the formation of D and R from DR) can be defined by $[DR] \times k_2$.

k_2 is a constant, the rate constant, of the reverse reaction.

It will be understood that both reactions are occurring simultaneously. As the reactions proceed an equilibrium is produced. This means that after some time the forward reaction and the reverse reaction are proceeding at the same speed and that the concentration of DR is constant. Once equilibrium has been reached we can write

$$k_1[D][R] = k_2[DR]$$

Simple algebra allows us to write

$$\frac{k_2}{k_1} = \frac{[D][R]}{[DR]}$$

We define the term k_2/k_1 as K_D, the dissociation constant for the reaction

$$K_D = \frac{[\text{D}][\text{R}]}{[\text{DR}]}$$

Now then, if we define the total number of receptors as R_T, the number of receptors that have, at equilibrium, reacted with D as DR and the number of free receptors (that is receptors which have not reacted with D) as R, we can see that

$$[R] = [R_T] - [DR]$$

We can now substitute $[R_T] - [DR]$ for $[R]$ in the equation for K_D

$$K_D = \frac{[D]([R_T]-[DR])}{[DR]}$$

$$K_D = \frac{[D][R_T]-[D][DR]}{[DR]}$$

$$K_D = \frac{[D][R_T]}{[DR]} - \frac{[D][DR]}{[DR]}$$

$$K_D = \frac{[D][R_T]}{[DR]} - [D]$$

$$K_D + [D] = \frac{[D][R_T]}{[DR]}$$

$$[DR](K_D + [D]) = [D][R_T]$$

$$\frac{[DR]}{[R_T]} = \frac{[D]}{K_D + [D]}$$

Let us now introduce a new term, r. We shall define r as the fraction of the receptors occupied by the drug. This is an important concept because it is likely that this proportion controls the response and r could be said to stand for the magnitude of the response.

$$\frac{[DR]}{[R_T]} = r$$

then

$$r = \frac{[D]}{K_D + [D]}$$

$$rK_D + r[D] = [D]$$

$$\frac{rK_D}{[D]} + r = 1$$

$$\frac{rK_D}{[D]} = 1 - r$$

$$\frac{rK_D}{1-r} = [D]$$

$$[D] = \frac{r}{1-r} K_D$$

This rather tedious algebra leads us to a graph. Imagine we plotted r on the y axis and $[D]$ on the x axis. First let us recast our last equation in a new form and do some more algebra

$$[D](1-r) = rK_D$$

$$[D] - [D]r = rK_D$$

$$\frac{[D]}{r} - [D] = K_D$$

$$\frac{[D]}{r} = K_D + [D]$$

$$[D] = r(K_D + [D])$$

$$r = \frac{[D]}{K_D + [D]}$$

$$r = \frac{1}{K_D + [D]}[D]$$

We can see that if we let r be y, and D be x then the gradient, m, is $1/(K_D + [D])$ and there is no value for C, the intercept on the y axis. The graph will, therefore, pass through zero but it will *not* be a straight line. Why not? Because the term for the gradient includes the value for x, *i.e.*, $[D]$. In fact the graph will be a curve. But what sort of curve? It will, in fact, be a hyperbola.

What we need to remember is that as $[D]$ increases, r will also increase but ever more slowly. In fact as $[D]$ goes on increasing the curve will become flatter and flatter until it runs almost parallel with the x axis. If we plotted the logarithm of $[D]$ on the x axis and, again, r on the y axis, we would obtain an S-shaped curve. This sort of curve is called a sigmoid curve. Now then, these are just the sort of curves obtained when the dose–response relationship for some toxic substance is plotted. Remember that we defined r as equivalent to the response. A large part of the sigmoid curve is nearly a straight line, a very useful feature when predicting the effects of increasing the dose of a drug.

The enthusiastic reader may be interested to know that a little more algebra can convert the sigmoid curve just mentioned into a straight line. Let us go back to an earlier equation

$$[D] = \frac{r}{1-r} K_D$$

By taking logarithms this equation can be converted into

$$\log[D] = \log\left(\frac{r}{1-r}\right) + \log K_D$$

or

$$\log\left(\frac{r}{1-r}\right) = \log[D] - \log K_D$$

If we plotted $\log(r/(1-r))$ on the y axis and $\log[D]$ on the x axis we would obtain a straight line. The familiar equation

$$y = mx + C$$

shows that in the above equation m, the gradient, is equal to 1.

To understand logistic regression, we need to understand the concepts of "odds". Let us start with an example from horseracing and how bookmakers assess the form of horses before a race. Assume a racehorse has won twice in its last 10 races. Based upon this form alone, what is the chance (probability) it will win its next race? The answer is 1 in 5 or 0.2 or 20%. But we also know it is more likely to lose than to win – it lost 8 times in its last 10 races. The probability of it losing the next race is 0.8 (4 in 5). An alternative way of expressing this information, favoured by bookmakers, is to use "odds". Odds relate the probability of an event occurring to the probability that the event does not happen. In this case, the odds of the horse winning are "1 to 4", also written "1 : 4". That is, in 5 races, the horse will win once and lose 4 times. Since losing is more common than winning, bookmakers usually quote odds against winning; "4 to 1 against" in this example. More formally, odds are defined as the probability of an event occurring (denoted p), divided by the probability that the event will not occur $(1 - p)$:

$$\text{Odds} = \frac{p}{1 - p}$$

Odds and probability convey similar information but on different scales; probabilities range from 0 to 1; odds from 0 to ∞. Table 12.1 shows the odds for a range of probabilities. For a probability very close to zero the odds are very close to zero; for a probability very close to 1 the corresponding odds are very large.

Let us now turn to a health-related example to develop our understanding of odds and how that leads us to logistic regression.

Outdoor air pollution has been classified as carcinogenic (causes cancer) by the International Agency for Research on Cancer (https://www.iarc.who.int/). As exposure to air pollution proceeds the cancer diagnosis, let A represent the presence of air pollution; $\bar{A}$ absence of air pollution; B cancer diagnosis; and $\bar{B}$ no cancer diagnosis. The odds that B will occur given the presence of A is

Table 12.1 Odds for a range of probabilities.

p	$1 - p$	Odds
0.01	0.99	0.01
0.1	0.9	0.11
0.2	0.8	0.25
0.4	0.6	0.67
0.6	0.4	1.50
0.8	0.2	4.00
0.9	0.1	9.00
0.99	0.01	99.00

$$\Omega_A = \frac{P(B|A)}{1-P(B|A)} = \frac{P(B|A)}{P(\bar{B}|A)} \tag{12.1}$$

Similarly, the odds that B will occur when A is not present is

$$\Omega_{\bar{A}} = \frac{P(B|\bar{A})}{1-P(B|\bar{A})} = \frac{P(B|\bar{A})}{P(\bar{B}|\bar{A})} \tag{12.2}$$

To assess whether the presence of A is associated with the occurrence of B we take the ratio of the two odds: (12.1) divided by (12.2). This gives the "odds ratio". An odds ratio (OR) above 1 tells us the odds that B will occur in the presence of A is greater than the odds that B will occur if A is absent; if the two odds are equal then the ratio equals 1 and if the OR <1 (but greater than 0 by definition) then the odds that B will occur in the presence of A is less than the odds that B will occur if A is absent.

We now have a way of assessing an association between two dichotomous variables. We can extend this to other explanatory variables, both categorical and continuous. The technique is called logistic regression and arises from a representation of the rate of occurrence of B assuming the presence of A (denoted C_A together with x covariates) as

$$P_A = \frac{1}{1+e^{-(mx+c_A)}} \tag{12.3}$$

The corresponding rate of occurrence of B assuming the absence of A is

$$P_{\bar{A}} = \frac{1}{1+e^{-(mx+c_{\bar{A}})}} \tag{12.4}$$

Dividing (12.3) by (12.4) and after some manipulation we arrive at the odds ratio

$$\frac{\Omega A}{\Omega \bar{A}} = e^{(c_A - c_{\bar{A}})} \tag{12.5}$$

Taking natural logarithms gives us

$$\ln \frac{\Omega A}{\Omega \bar{A}} = c_A - c_{\bar{A}} \tag{12.6}$$

Hence log(OR) is the difference between the two parameters that define exposure.

More generally (12.6) is written:

$$\ln\frac{\Omega A}{\Omega \bar{A}} = a + b_1 x_1 + b_2 x_2 \ldots + b_n x_n \tag{12.7}$$

Hence ln(OR) is expressed as a linear function of explanatory variables with the parameters, β_i, describing the rate of change in ln(OR) per unit change in x_i. It follows then that the OR for a unit change in x is simply e^{β}. The α term in eqn (12.7) represents the base line odds. The logistic model (12.7) is commonly used for prediction and classification problems and has many applications including in medicine, social science, finance and machine learning.

We can now model, for example, the probability of a cancer diagnosis given exposure to air pollution. Figure 12.7 shows the odds ratio for a cancer diagnosis associated with the pollutant concentration, x (NB these are artificially generated data for illustration only). It is important to note that individual patients either do, or do not, have the disease. We are interested in the *probability* of the outcome not the actual outcome for individual patients. The curve in Figure 12.7 derived from the logistic model is the sigmoid or logistic curve.

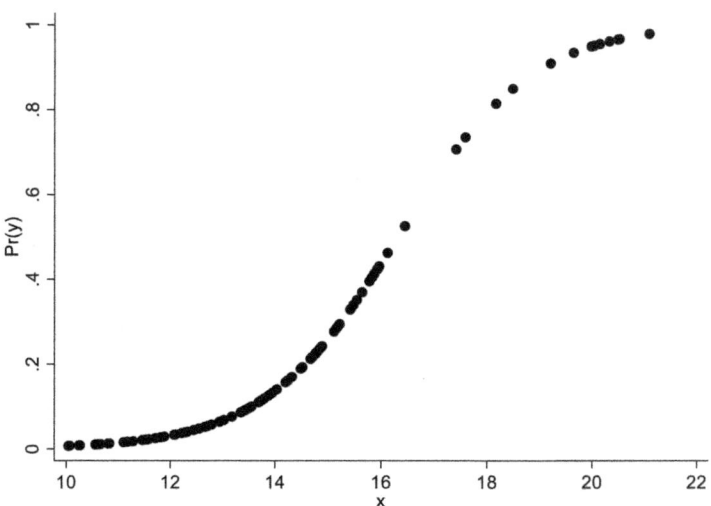

Figure 12.7 Hypothetical data demonstrating the relationship between probability of a cancer diagnosis (*y*-axis) given exposure to air pollution (*x*-axis).

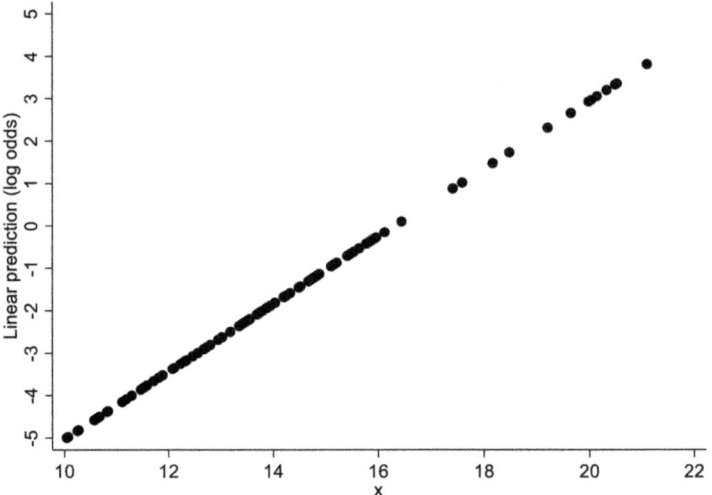

Figure 12.8 Hypothetical data demonstrating the relationship between ln(odds) of a cancer diagnosis (*y* axis) given exposure to air pollution (*x* axis).

Plotting the ln(OR) shows, as expected from (12.7), a straight line relationship with the pollutant concentration – Figure 12.8.

We return to odds ratios in Chapter 13 when we introduce study designs used in air pollution epidemiology.

12.9   Poisson Regression

Work on the effects of air pollutants on health often involves the use of Poisson regression: it is the standard method used in time-series studies; see Chapter 13. The results of a Poisson regression are rather like those of a logistic regression: the independent variable, plotted on the *x* axis, is plotted on an arithmetic scale; the dependant variable, the odds ratio in logistic regression and the relative risk in Poisson regression is plotted on the *y* axis. The following equations show the similarity: in logistic regression

$$OR = e^{\beta}$$

In Poisson regression

$$RR = e^{\beta}$$

In fact, when events are rare the odds ratio can be interpreted as a relative risk.

The methods are also similar in that you cannot manage with a simple calculator: again an iterative process is used. Where the methods differ is that logistic regression focuses on whether an event occurs or not, Poisson regression focuses on how often a rare or infrequent event occurs. In all the regression methods that we have discussed there is a buried distribution curve: the normal distribution in ordinary linear regression, the binomial distribution in logistic regression, the Poisson distribution in Poisson regression. The assumption that one of these distributions applies to the data means that these methods are parametric methods.

12.10   Non-parametric Regression

Those readers who have come across both parametric and non-parametric statistical tests will recall that the former requires us to be able to define a distribution curve: normal, Poisson, Student's *t* distribution, binomial, *etc*. Non-parametric methods require no such decision about a distribution, they are described as distribution-free methods. Such methods were not included in the textbooks of statistics of many years ago and were, rather disparagingly, referred to as "quick methods" but are now widely used. Some readers will have come across Wilcoxon's tests and the Mann and Whitney tests, perhaps even the Kolmogorov–Smirnov tests; these can be used instead of the more familiar "Student's *t* test" and are often more satisfactory. There are statistical techniques designed to tell us, on the basis of the distribution of the data, whether a parametric test is appropriate, if not: use a non-parametric test. Without labouring the point let us accept that there are also non-parametric regression techniques.

The essence of non-parametric regression is that no fundamental assumption regarding the shape of the regression line is required. Put simply, the line follows the data but the method includes mathematical "dodges", which are applied to produce a smooth curve without too many sharp bends: sharp bends are regarded as artefacts and are avoided! How simple that sounds; how difficult it is to understand in detail! Such methods are widely used in modern air pollution epidemiology and research papers are full of details of splines and knots: these are some of the "dodges". A basic moving average is perhaps the simplest and easiest non-parametric smoothing method and a good concept to start from. A moving average, is as the name suggests an average (of a set of values) that moves through the data. It is best illustrated by an example. Consider the following series of values taken on consecutive days, $t = 1$ to 8: 2, 6, 4, 5, 12, 4, 6, 7. A three-day moving average of the series would give values of 4: $(2 + 6 + 4)/3$, 5: $(6 + 4 + 5)/3$, 7: $(4 + 5 + 12)/3$, *etc.* The scope of this text is too limited to attempt a satisfactory explanation of the details of the methods. However, let us look at the sort of curves which can be produced. The following examples (Figure 12.9) are taken, with permission, from Thurston *et al.*[1]

Let us look at the curves. First the axes: on the x axis annual average $PM_{2.5}$; on the y axis: the mortality hazard ratio, rather like the relative risk.

Just above the x axis the black bars which become a series of short vertical lines at each end, shows the distribution of the data: very low and very high concentrations were rare. The short vertical lines which make up this illustration of the distribution of the data comprise a "rug plot": it looks a bit like a sideways view of a rug which is becoming tattered at each end. Now for the regression line. Obviously, in neither illustration is it a straight line. Indeed it was not constrained to be a straight line: in fact the line follows the data; well, it does up to a point. The lines are in fact estimated using natural spline models: we have already mentioned these as mathematical dodges used in fitting non-linear curves.

The dotted lines are the 95% confidence intervals. For what? For the position of the regression line. Note that at the ends of the regression line, where the data are sparse, the confidence intervals open up like the mouths of trumpets. Near the centre of the regression line,

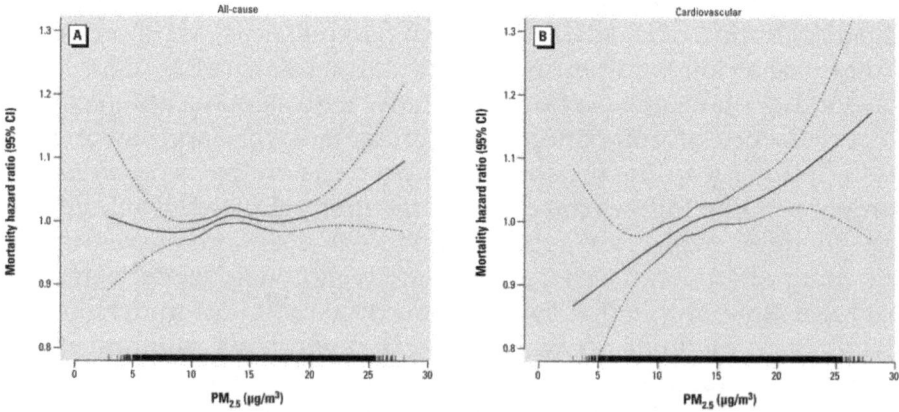

Figure 12.9 Concentration response curves and 95% confidence intervals. Reproduced from ref. 1.

where data is plentiful, the confidence intervals are rather close to the regression line. We may interpret the 95% confidence interval lines as meaning that in 95% of cases the confidence interval will include the true position of the line. This means that we should not place too much reliance on the position of the ends of the regression line. Near the midpoint of the regression line we should have much more confidence in its position.

Now for some really difficult questions.

What do we make of the shape of the regression lines? We are not bound to make anything of it: the line is the line, it describes the relationship. But why is it the shape it is? What mechanisms underlie the shape? Why, in the left-hand panel does it turn upwards at low concentrations of fine particles and why, in the right panel, does it keep going down? Why, in some work, is a supra-linear relationship between concentration and effect at low concentrations reported? At present there are no answers to these questions, at least no mechanistic answers that appeal strongly to a toxicologist.

One last point.

Regression techniques tell us about relationships between independent and dependant variables. Regression techniques do not explain those relationships. Regression techniques do, within limits, allow us to predict the effects on the dependant variable of changes in the independent variable. But remember, all models are wrong, but some are useful!

References

1. G. D. Thurston, J. Ahn, K. R. Cromar and Y. Shao, *et al.*, Ambient particulate matter air pollution exposure and mortality in the NIH-AARP diet and health cohort, *Environ. Health Perspect.*, 2016, **124**, 484.

13   Air Pollution and Health

13.1   Introduction

The health effects of exposure to outdoor air pollution have been studied extensively. One of the earliest studies was an assessment of the notorious London "smog" episode of December 1952 in which an estimated 3500 to 4000 extra deaths were observed compared to what would have been expected for that time of year.[1] The introduction of the Clean Air Acts in the 1950s and 1960s led to a reduction in concentrations of particulate matter (Black Smoke) and sulfur dioxide produced by combustion of coal. Interest in air pollution, from the health perspective, then dwindled. However, with the subsequent growth of traffic, and traffic pollution, the health effects of exposure to outdoor air pollution became a concern once again. Two early studies, published in the 1990s, serve to illustrate the principal approaches taken to the study of air pollution and health and we shall discuss these further later in this chapter.

There now exists a vast literature examining possible adverse effects of outdoor air pollution on a range of diseases including respiratory and cardiovascular diseases, cancers, neurological conditions and foetal development. Health outcomes studied have included symptom (coughing, wheezing, shortness of breath, *etc.*) frequency and prevalence (the proportion of the population with a health condition/disease), medication use, primary care (GP) consultations, visits to emergency (A&E) departments, emergency hospital admissions and death. Air pollutants studied include particulate matter (mass, size and composition) and gaseous pollutants (*e.g.* ozone, nitrogen dioxide

Basic Mathematics for Students of Air Pollutants
By Robert Maynard and Richard Atkinson
© Robert Maynard and Richard Atkinson 2024
Published by the Royal Society of Chemistry, www.rsc.org

and carbon monoxide). Concentrations of air pollutants used in studies have come from routine and campaign monitoring networks and from modelled air pollutant concentrations (a large field in its own right), informed, and validated, by monitored data. The evidence provided by the large number of studies worldwide has guided air pollution abatement strategies and public policy.

The purpose of this chapter is to introduce some key concepts and terms used in air pollution and health research such that the results of studies can be understood and interpreted appropriately. We also give a brief introduction to the topic of systematic review and evidence synthesis – processes designed to identify relevant literature, assess its quality and combine results using graphical and numerical methods such that the totality of the evidence can be evaluated and summarised. Systematic reviews are now commonplace in medical research and form the evidence base for policy development. Numerous examples in air pollution and health research exist and we shall refer readers to some important reviews.

Epidemiology is "the study of how often diseases occur in different groups of people and why".[2] In air pollution epidemiology we examine the hypothesis that the "why" is exposure to air pollutants. Epidemiology is based upon observational study – examining patterns of disease and determinants or risk factors in populations and, hence, it is the study of observed associations. The question of whether exposure "A" causes disease "B" is another matter, and one that requires consideration of a number of other factors to determine if an association is causal. We do not propose to discuss this in any detail but have provided an introduction to the problem in Chapter 14. For more information the interested reader is referred to the informative texts by Bradford Hill,[3] Rothman and Greenland[4] and Cox.[5] By comparison, in a clinical trial, participants are assigned randomly to either the group receiving the intervention (*e.g.* a new treatment) or the control group (*e.g.* standard treatment). The randomisation ensures the groups are comparable in all respects, other than the treatment received, and differences in the outcome at the end of the trial are therefore attributed to (caused by) the intervention.

13.2 Study Designs

Air pollution epidemiology divides broadly into two categories: studies of air pollution exposure over the long-term (measured in years) and studies of short-term exposure (measured in hours or days).

Examples of the former include cohort and case control studies, and of the latter, time-series and panel studies. As indicated earlier, we will consider further two seminal studies, both from the 1990s; one a cohort study, the other, a time-series study, and use each to illustrate the main features of their respective designs.

13.2.1 Cohort Studies

A cohort study is a prospective, longitudinal study: it starts with participants without the disease or outcome having occurred. Their exposures to the relevant risk factors are ascertained at the start of the study and the participants are followed up over time until the outcome of interest (first) occurs or the study ends. The American Cancer Society Cancer Prevention Study II (ACS CPS-II) cohort was first published in 1995.[6] Since then there have been numerous extended analyses and reanalyses of this cohort that have demonstrated that long-term exposure to fine particulate matter air pollution ($PM_{2.5}$) and other pollutants is associated with increased risk of mortality (by definition, mortality is an incident event). In the original study, participants (552 000) were friends or family members of cancer patients from across the US recruited in 1980 and followed over 7 years (1982–1989) to determine survival. Participant exposures to air pollutants were *estimated* by the mean or median annual concentrations of sulphate and fine particulate matter (from monitoring networks) for the metropolitan area in which each resident lived (hence all participants living in the same metropolitan area were assigned the same exposure concentrations). The relationship between assigned air pollution concentrations and risk of death during the follow-up period is expressed as a hazard ratio and was estimated using a statistical method called survival analysis. We will not discuss the detail of the method but for the interested reader there are numerous excellent texts on the subject (see Further Reading).

At this point we need to introduce a new concept – confounding. Confounding is important in observational studies as its presence, and failure to take account of it in the analysis, can lead to false conclusions. It is best explained by an example. It is well established that smoking tobacco causes cancer, respiratory and cardiovascular disease leading to premature death. Let us assume air pollution is higher in areas of the country where more people smoke and lower in areas of the country where fewer people smoke. A study of air pollution and death that did not account for smoking rates might well report an adverse association between air pollution concentrations and premature death. It is entirely feasible that this observed association may, in fact, be due to smoking, and that failure to account for smoking rates in the analysis has resulted in a spurious association with air pollution. Smoking is a confounder of the association between air pollution and premature death. How can we deal with this problem? The answer is to include smoking in the analysis so that the association between air pollution and premature death is adjusted for the effects of smoking. Confounding, due to smoking and many other factors (age, diet, physical exercise, weather, *etc.*), is a key element in air pollution epidemiology and one the reader should be aware of when assessing evidence from these studies. It is important to note that only confounders that vary on the same scale as concentrations of pollutants and appropriate to the study design are relevant. For example, a confounder that is uniform spatially but that varies on a day-to-day basis (such as daily weather conditions) is not relevant in a cohort design. Later in this chapter we will take a brief look at epidemiological studies that assess associations between air pollution and health on a day-to-day basis – daily weather conditions are very important confounders in such studies but factors that vary little over time, such as smoking status, are not. Cohort studies collect therefore not only information on air pollution, date and cause of death but also information on many potential confounders including the characteristics and lifestyle of individuals in the study, demography, and other environmental risk factors. These factors are combined in a statistical model and the association between the air pollutant concentration and premature death estimated.

What do the results of these analyses look like? The association between air pollution concentration and death is captured by a statistic called a hazard ratio (HR). We will not go into the statistical theory underpinning its estimation. It is often interpreted as a relative risk. It is worth describing the relative risk in a little more detail because it is commonly reported in a range of study designs in air pollution epidemiology.

The relative risk (RR) is the ratio of two risks: the risk of an event *e.g.* death at one level of a risk factor divided by the risk of the event at another level of the risk factor. For example, risk of premature death in smokers *versus* risk of premature death in non-smokers. For air pollutants measured on a continuous scale the risk of an event at, or associated with, one pollutant concentration is divided by the risk at another (lower) pollutant concentration. If the two risks are the same, the RR = 1. If the risk of death is greater at higher, compared to lower, pollutant concentrations then the RR > 1. If the estimated risk is lower at higher pollutant concentrations, compared to lower pollutant concentrations then, of course, the RR < 1.

Returning to the ACS CPSII study, the HR for $PM_{2.5}$ and all-cause mortality was given as 1.17 (95% CI: 1.09–1.26) per 24.5 µg m^{-3} (the difference between mean $PM_{2.5}$ concentrations in the most *vs.* the least polluted areas). How do we interpret the 1.17? We say the risk of death is 1.17 times (or 17% higher) in an area where mean $PM_{2.5}$ concentrations are 24.5 µg m^{-3} greater than another area in the study. The confidence interval indicates the precision of that estimate – the true increase in risk per 24.5 µg m^{-3} may be as low as 1.09 or as high as 1.26. Note, this estimate was adjusted for age, sex, smoking status, race, education, occupational exposure, body mass index (BMI) and alcohol consumption.

There are a number of important features of the RR to appreciate. First, the relative risk is, as the name suggests, a relative measure – it quantifies the ratio between two risks. It does not tell us anything about the actual sizes of the risks. A 17% increase in the risk of a rare event has quite different implications compared to a 17% increase in the risk of a common event. For example, a Swedish study[7] reported that women who attended university were 23% more likely (RR = 1.23) to develop a type of cancerous brain tumour compared with women who did not attend university. A worrying, if somewhat perplexing, finding. The incidence rate of gliomas in the UK is approximately 2 per 100 000 persons.[8] The 23% increase, *if causal*, raises the risk to ~2.5 per 100 000 – a much less alarming perspective on the findings of the study!

Secondly, the magnitude of the RR depends not only upon the strength of the relationship between the risk factor and outcome but also upon the increment in the risk factor used to calculate the RR. For example, a RR of death of 1.05 associated with a 2 µg m^{-3} increment in pollutant concentration equates to a RR of 1.28 per 10 µg m^{-3}. Changing the pollutant increment used changes the magnitude of the RR. One can easily emphasise a positive association through judicious selection of concentration change. At first glance, one might expect the RR for a 10 µg m^{-3} increment to be 5× the risk at 2 µg m^{-3} increment. The reason it is not is because the statistical technique used estimates the association between air pollution and health on the log scale. Exponentiation gives the RR. How do you rescale a RR from one pollutant increment to another? The simplest way to approach this is to first change to the log scale, rescale ln(RR) to a single unit pollutant increment, then multiply by the new increment before exponentiating. Mathematically this rescaling is expressed as follows:

$$\mathrm{RR}\,|\,P1 = \exp\left(P1 * \left(\frac{\ln(\mathrm{RR}\,|\,P0)}{P0}\right)\right)$$

where $P0$ and $P1$ are respectively the initial and new pollutant increments and "|" means "given".

Thirdly, RR is an example of an additive relationship on the log scale and multiplicative on the linear scale. Hence, an increase in pollutant concentrations from 10 μg m^{-3} to 20 μg m^{-3} gives the same *relative* increase in risk as an increase from 110 μg m^{-3} to 120 μg m^{-3}. The multiplicative relationship (for a given pollutant increment) is constant across the range of pollutant concentrations. In formulating public health policy, mitigation strategies aimed at reducing pollution concentrations are evaluated against a counter-factual (alternative) pollutant concentration. Ignoring non-anthropogenic sources for now, we can work out risk estimates (y) at different values of a pollutant (x), each relative to zero concentration ($x = 0$). As an example, let's take a typical result for PM$_{2.5}$ and mortality: a relative risk of 1.006 per μg m^{-3} increment in PM$_{2.5}$. Let us also tabulate the natural logarithms of y. Remember that we are not looking at "real data", we are looking at the predictions of the model used to describe "real data". Table 13.1

Table 13.1 Relative risk (y) for different concentrations of pollutant (x) relative to zero concentration (relative risk of 1.006 per μg m^{-3} increment in PM2.5).

x	y	$\ln(y)$
0	1.00	0
10	1.06	0.0598
20	1.13	0.1196
30	1.20	0.1795
40	1.27	0.2393
50	1.35	0.2991
60	1.43	0.3589
70	1.52	0.4187
80	1.61	0.4786
90	1.71	0.5384
100	1.82	0.5982
200	3.31	1.1964
400	10.94	2.3928
600	36.21	3.5892
800	119.78	4.7857
1000	396.26	5.9821

shows the relative risk, *y*, for different values of *x*, each relative to zero concentration.

The data are presented graphically in two parts: Figure 13.1 presents data for concentrations of *x* from 0 to 100 µg m³; and Figure 13.2 for values of *x* from 200 µg m⁻³ to 1000 µg m⁻³. It will be immediately seen that the plots of ln(*y*) against *x* are straight lines, whereas the plots of *y* against *x* are curves. Note from Figure 13.2 that as *x* exceeds about 600 µg m⁻³ the line begins to "take off". The figures illustrate the multiplicative nature of the relationship with much steeper increases observed in *y* for larger values of *x*. The multiplicative risks are compounded as *x* increase *compared to the counterfactual*.

Finally, if we choose a value of *x*, say 100, and compare the effect on *y* of increasing *x* by 10 (from 100 to 110) and of decreasing *x* by 10 (from 100 to 90), we can see that the increase in *x* produces a larger change in *y* than does the decrease in *x*. Why is this? The important point to remember is that we are dealing with proportional changes. The RR for a 10 unit increase in *x* *is* 1.06 or a 6% increase. The equivalent decrease is a relative risk of 1/1.06 = 0.943, or a 5.7% decrease in risk. A slightly more obvious example would be: a 50% increase on 100 is 150. But a smaller % decrease is required to change from 150 to 100 as the starting point is larger: (150 − 100)/150 = 0.33 or 33% decrease.

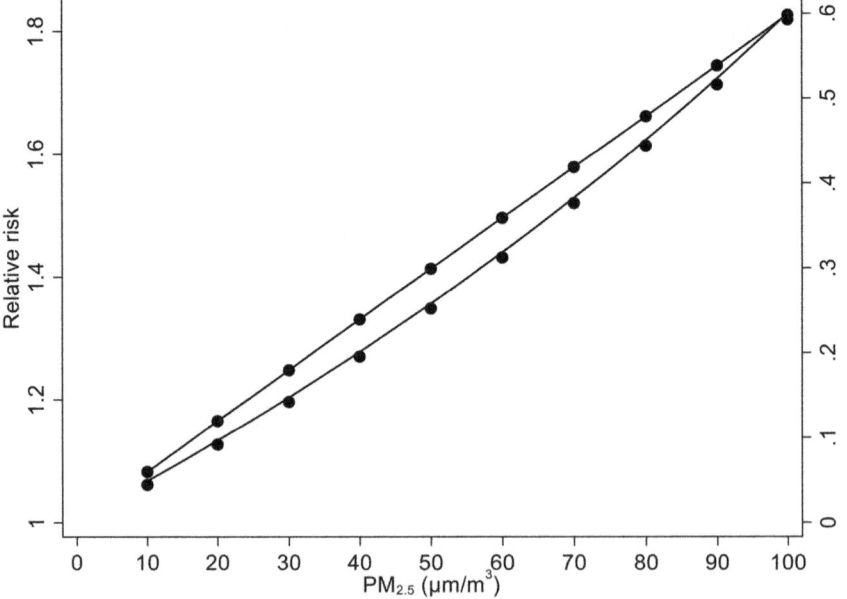

Figure 13.1 Plot of relative risk and log relative risk against concentration of PM₂.₅ between 0 and 100 µg m⁻³.

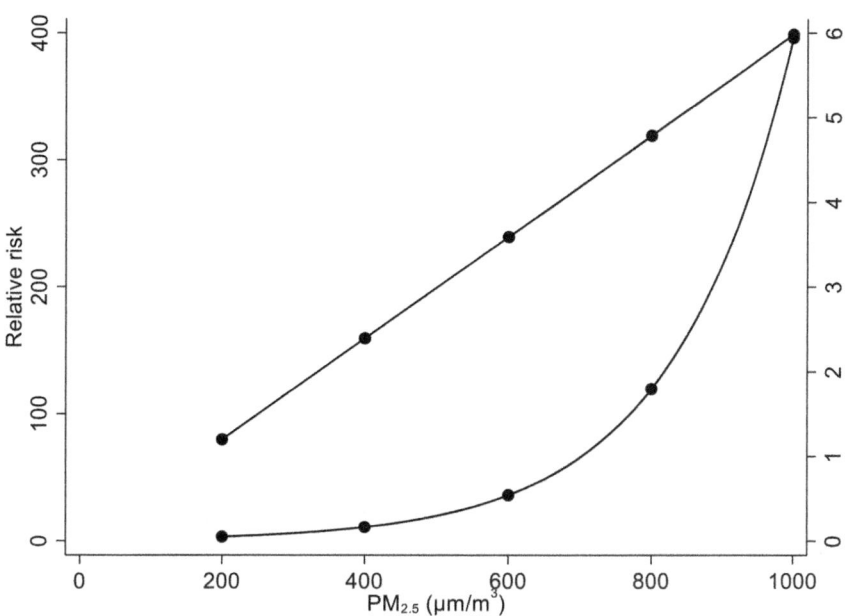

Figure 13.2 Plot of relative risk and log relative risk against concentration of $PM_{2.5}$ between 0 and 400 µg m^{-3}.

The equation to calculate the relative change in risk associated with an increase in concentration from $x0$ to x is a simple one:

$$y = 1.006^{x-x0}$$

To calculate the *absolute* change in risk between two concentrations one must first calculate both risk estimates and then subtract. Let us assume a starting point of $x = 50$ and calculate the effects, in absolute terms, on y of an increase in x of 10.

$$y_{x=50} = 1.006^{50-0} = 1.3486$$

Calculating 1.006^{50} is easy with a calculator. Alternatively, take the log of 1.006, multiply by 50 and take the antilog of the product.
Similarly:

$$y_{x=60} = 1.006^{60-0} = 1.4318$$

$$y_{x=40} = 1.006^{40-0} = 1.2703$$

$$y_{x=60} - y_{x=50} = 1.4318 - 1.3486 = 0.0832$$

$$y_{x=50} - y_{x=40} = 1.3486 - 1.2703 = 0.0783$$

There *is* a small difference, the line *is* curved.

Let us now consider the question of how to compare RR for different pollutants. Imagine that, in a cohort study, we could estimate separately the effects of three pollutants: PM monitored as PM_{10}, sulfur dioxide and nitrogen dioxide, and that the coefficients describing the regression lines were: 0.72%, 0.6% and 0.54%. Comparison of these coefficients would tell us that, on a "per $\mu g\ m^{-3}$" basis, PM monitored as PM_{10} had a greater effect than sulfur dioxide and that sulfur dioxide had a greater effect than nitrogen dioxide.

Let us now express the coefficients in terms of the inter quartile range (IQR), the difference between the 75th and 25th percentile values of the distribution for the three pollutants. The reader will remember that we discussed quantiles in Chapter 5. Imagine that the coefficients, expressed as percentage change in RR per IQR, were 4.3, 6.9 and 8.2 per $\mu g\ m^{-3}$ for PM monitored as PM_{10}, sulfur dioxide and nitrogen dioxide respectively. Let us condense the data into a table (Table 13.2).

We can now make two statements:

1. In terms of the variability of the concentrations of the three pollutants the effect of nitrogen dioxide is greater, over a representative range of exposures, than the effect of sulfur dioxide, and the effect of sulfur dioxide is greater than that of PM monitored as PM_{10}. The concentration of nitrogen dioxide varies much more than that of PM monitored as PM_{10} and thus has a more important effect on health.
2. In terms of the toxicological activity of the pollutants the toxicity of PM monitored as PM_{10} is greater than that of sulfur dioxide, and the toxicity of sulfur dioxide is greater than that of nitrogen dioxide.

Or, put another way, coefficients expressed in terms of IQR tell us about the relative importance to health of pollutants in the locations studied; coefficients expressed in terms of "per unit $\mu g\ m^{-3}$" tell us about the relative toxicity of pollutants.

Table 13.2 Relative risks expressed for different pollutant increments.

	PM_{10}	SO_2	NO_2
IQR ($\mu g\ m^{-3}$)	6	11.6	15.2
% change in RR per IQR	4.3	6.9	8.2
% change in RR per $\mu g\ m^{-3}$	0.72	0.6	0.54

It will be clear that the relative importance to health of a series of pollutants might well vary from location to location because the IQR for the pollutants varies from location to location. One might think that the coefficients expressed in terms of $\mu g\ m^{-3}$ should be stable from location to location. For the gases sulfur dioxide and nitrogen dioxide this should be the case. But for PM monitored, for example, as PM_{10} this would only be the case if the compositions of PM_{10} were identical from location to location: this is not so. We have made another assumption: that the pollutants do not interact in a toxicological sense. Of course, they might interact and this interaction might well depend on the relative concentrations of the individual pollutants and on the compositions of PM_{10}. We should, therefore, not be surprised if the coefficients, expressed on a per $\mu g\ m^{-3}$ basis, for sulfur dioxide and nitrogen dioxide, as well as that for PM monitored as PM_{10} varied from location to location.

Care is needed in interpreting coefficients: always examine the units.

13.2.2 Other Study Designs for Investigating Relationships Between Long-term Exposure to Air Pollutants and Effects on Health

Let us leave the nuances of risk and relative risk and conclude the discussion of epidemiological study designs by mentioning two other study designs used to assess associations between long-term concentrations of air pollutants and health: case-control and cross-sectional studies.

Case-control studies are retrospective observational studies that aim to compare the presence of potential determinants of disease in subjects with the disease of interest (cases) with subjects without the disease (controls). It is a study design that is often used in the study of rare diseases. A cohort study would be inappropriate for a rare disease as a very large number of subjects would be required for sufficient numbers of subjects to go on to develop the disease. The major challenge in case-control studies is finding suitable controls. If a rare disease is usually found in elderly people then selecting young people for controls would not make sense – they are not a comparable group for a valid comparison. The answer is to use matching; selecting controls so that they match on key potential confounders, often age and sex, and to try to select controls from the same population. Patients admitted to hospital are unlikely to be similar to patients not in hospital!

In simple case-control studies the outcome is binary – subjects are either cases with disease or controls without disease and the relationship with explanatory variables is of interest. This brings us back to logistic regression introduced in Chapter 12. Analysis of case-controls studies gives us an odds ratio comparing odds of the presence of a factor or determinant of disease in cases compared to the odds of the presence of the factor in controls. In case-control studies we start with the outcome, the subject has the disease or not, and look back to see what their exposure to various factors was.

Before we leave case-control studies a brief note about the interpretation of odds ratios, and in particular, a comparison with relative risk. In case-control studies the OR is derived from examination of the odds of (past) exposure in patients with and without the outcome of interest. We must be careful not to simply interpret the OR as a description of a patients risk of developing disease in the future if they are exposed to a determinant of disease. Such a switch, from looking back to looking forward and assessing the risk of developing disease is reasonable if the disease in question is rare. The OR is then a good *approximation* to the relative risk (RR). A little algebra illustrates the relationship.

Let $P1$ represent the baseline proportion of patients experiencing an event and $P2$ the proportion experiencing the event following some intervention or exposure. The RR is defined as

$$\text{RR} = \frac{P2}{P1}$$

and the OR as

$$\text{OR} = \frac{P2/1-P2}{P1/1-P1} = \frac{P2(1-P1)}{P1(1-P2)} = \text{RR} * \left(\frac{1-P1}{1-P2}\right).$$

Table 13.3 illustrates the relationship between the RR (in this example set at 1.5) and the OR for different values of $P1$. When the baseline risk is low, the two ratios are close but soon deviate once the baseline risk increases. The OR is approximately double the RR when $P1$ is 0.2.

A brief word about cross-sectional studies. Cross-sectional studies describe disease (and risk factor) patterns in population samples. We will not consider these further but refer the reader to an overview of the evidence published by the US Health Effects Institute in 2022[9] which summarises these studies, as well as evidence from cohort studies.

Table 13.3 Odds ratios (OR) for different values of $P1$, the proportion of patients experiencing an event and $P2$, the proportion experiencing the event following an intervention or exposure.

$P1$	$P2$	OR
0.01	0.015	1.51
0.11	0.165	1.60
0.21	0.315	1.73
0.31	0.465	1.93
0.41	0.615	2.30
0.51	0.765	3.13
0.61	0.915	6.88

13.2.3 Time-series Studies

The second example we will consider in a little more detail is an example of a time-series study of air pollution and daily mortality in London published in 1996.[10] The time-series study design aims to estimate associations between daily concentrations of pollutants across a city or geographical area and daily counts, across the same area/location, of health events. Such events may be deaths from specific causes, hospital admissions or other population-based measures of disease. The design also enables the time lag (1 day, 2 days, *etc.*) between exposure and health events to be investigated. The associations of interest in a time-series study are temporal (the unit of analysis is the day), rather than spatial (where the unit of analysis is the person) as in a cohort study.

As with other observational air pollution studies control for potential confounders is very important. Individual characteristics, such as smoking status, BMI, *etc.* do not vary on a day to day basis and so are not considered confounders in the analysis. However, temporal factors that do vary over the study time period and can play an important role in determining the number of daily health events are potential confounders. These include season, daily temperature and humidity and seasonal diseases such as influenza. The statistical methods used in time series study account for these factors to reveal the nature of short-term associations between air pollution concentrations and health events. And whilst the statistical method differs from those employed in cohort studies the output is, again, a relative risk. In the study by Anderson and colleagues[10] daily average concentrations of Black Smoke across London during the summer months were associated with an increased risk of death the following day, RR = 1.025 (95% CI: 1.009–1.041) for a 12 μg m^{-3} increase in BS. A very large number of time-series studies have been published investigating a range of pollutants, many diseases and health endpoints and in a large number of diverse locations. The interested reader is referred to a recent assessment of the evidence conducted by the World Health Organisation.[9]

13.3   Systematic Reviews and Evidence Synthesis

Systematic review and evidence synthesis is a relatively recent discipline. In his seminal text "Effectiveness and efficiency"[11] the epidemiologist Professor Archie Cochrane urged health practitioners to practice evidence-based medicine. Later, in his contribution to the "Medicines for the year 2000" symposium held at the Royal College of Physicians, Cochrane wrote: "It is surely a great criticism of our profession that we have not organised a critical summary, by specialty or subspecialty, adapted periodically, of all relevant randomised controlled trials."[12] The work of Cochrane and colleagues led to the establishment of the Cochrane Collaboration (https://community.cochrane.org/). As well as providing a large number of thorough, state of the art reviews on a wide range of health topics, the collaboration provides an excellent description of the systematic review and evidence synthesis methodology. Whilst the original focus of the collaboration and others was summarising results from randomised control trials (RCTs), the need to assess epidemiological evidence to support policy development led to the application/development of methods for observational studies, including those in the field of air pollution health research. Today, there are a large number of reviews and the interested reader is directed to the work of the WHO (https://www.who.int/), the US Health Effects Institute (HEI) (https://www.healtheffects.org/), Environmental Protection Agency (EPA) (https://www.epa.gov/) and the UK Committee on the Medical Effects of Air Pollutants (COMEAP). The Prospero website (https://www.crd.york.ac.uk/prospero/) maintains a database of registered systematic reviews, planned, in progress and completed. A search (May 2022) for reviews of the health effects of air pollution returned 491 systematic reviews! Perhaps not surprisingly, reviews of reviews are now being published.

   Systematic review and evidence synthesis is a large topic and we will not cover this in detail here. Instead we provide a brief outline of the methodology and include an example of the typical output found in air pollution reviews.

The review methodology can be thought of as two separate phases: (1) identification of the relevant literature, and (2) analysis of the study findings.

Identification of the literature involves defining suitable search terms to submit to the search engines of online medical database such as PubMed (https://pubmed.ncbi.nlm.nih.gov/) and Web of Science (https:// clarivate.com/products/scientific-and-academic-research/research-dis-covery-and-workflow-solutions/webofscience-platform/). Typical search terms include the pollutants of interest, the health outcome(s) and study designs. Searches should be inclusive such that no relevant studies are missed, but specific enough so as not to include a large number of irrele-vant studies. Searches are usually applied to a number of literature data-bases and the results combined and duplicates removed. The remaining studies are then assessed for eligibility for inclusion in the review first by screening study titles and abstracts and then by full text review. Searches can return potentially many thousands of relevant studies, so one can perhaps appreciate the huge amount of work that this process can entail. The process is summarised in a Prisma flow diagram (http:// prisma-statement.org/prismastatement/flowdiagram.aspx). An example from a recent WHO review is reproduced in Figure 13.3.[13]

Next begins the analysis stage. This involves extracting all relevant study information, including effect estimates, exposure informa-tion, and details of confounder adjustment. Usually these are (dou-ble) entered into an electronic database or spreadsheet to facilitate graphical and numerical analysis. Effect estimates from different studies are then standardised (to the same pollutant units and incre-ment so that they can be compared meaningfully across studies). The standardised study estimates can then be presented graphically in a forest plot, Figure 13.4.[13] The plot provides important information about the evidence base: the number of studies contributing evidence, the direction, size and precision of the individual study estimates and the variability between studies. Estimates can also be combined in a meta-analysis, the purpose of which is to provide a single summary estimate that, by combining studies, has greater precision than indi-vidual studies and facilitates a statistical assessment of the variability between study estimates (heterogeneity). There are two main methods for deriving the summary estimate: fixed-effects and random-effects models. Further exploration of these two models is beyond the scope of this book and we refer the reader to the excellent online Cochrane manual on systematic reviews (https://training.cochrane.org/hand-book/current). Figure 13.4 also shows the summary estimate and associated 95% confidence interval derived from the meta-analysis,

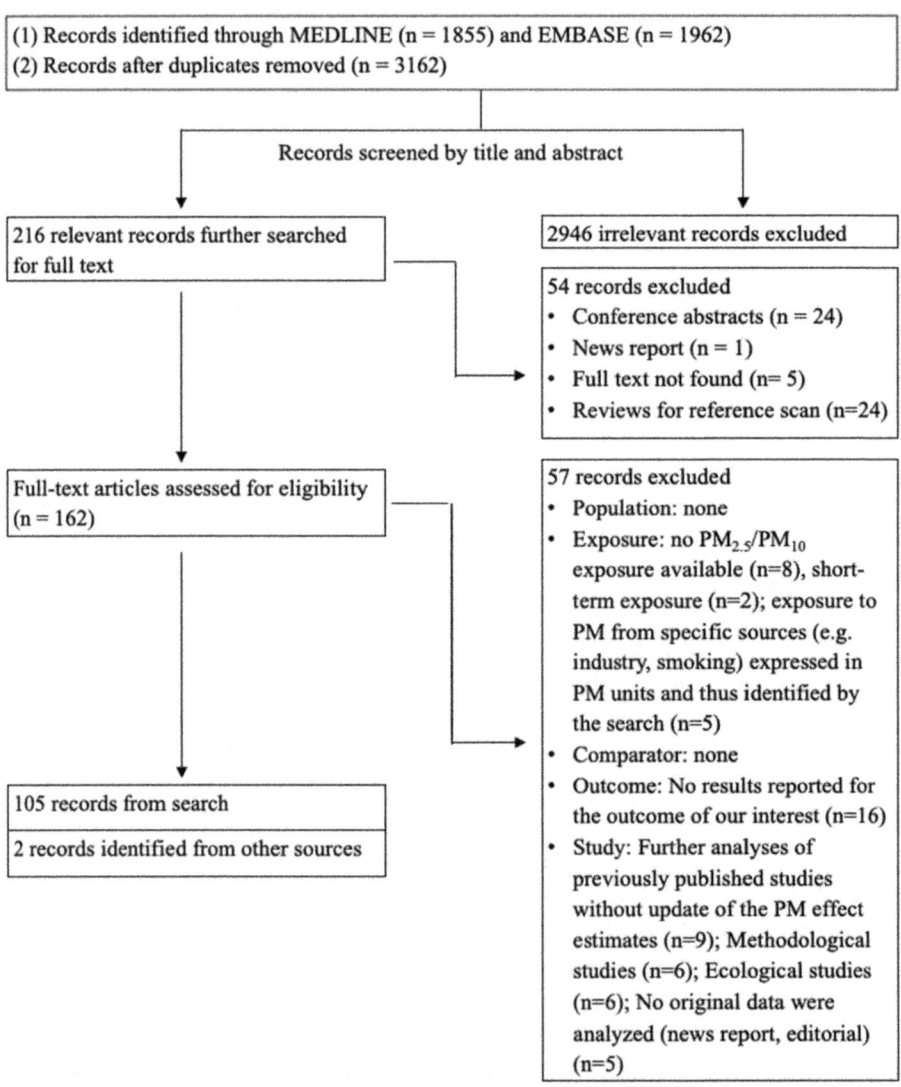

(1) Records identified through MEDLINE (n = 1855) and EMBASE (n = 1962)
(2) Records after duplicates removed (n = 3162)

Records screened by title and abstract

216 relevant records further searched
for full text

2946 irrelevant records excluded

54 records excluded
• Conference abstracts (n = 24)
• News report (n = 1)
• Full text not found (n= 5)
• Reviews for reference scan (n=24)

Full-text articles assessed for eligibility
(n = 162)

57 records excluded
• Population: none
• Exposure: no $PM_{2.5}/PM_{10}$
  exposure available (n=8), short-
  term exposure (n=2); exposure to
  PM from specific sources (e.g.
  industry, smoking) expressed in
  PM units and thus identified by
  the search (n=5)
• Comparator: none
• Outcome: No results reported for
  the outcome of our interest (n=16)
• Study: Further analyses of
  previously published studies
  without update of the PM effect
  estimates (n=9); Methodological
  studies (n=6); Ecological studies
  (n=6); No original data were
  analyzed (news report, editorial)
  (n=5)

105 records from search

2 records identified from other sources

Figure 13.3 Example of a PRISMA flow diagram. Reproduced from ref. 13 with permission from Elsevier, Copyright 2020.

here represented by the solid diamond and protruding solid line (row marked "RE model") with the actual values presented in the column on the far right of the figure. At the bottom left of the figure, various statistics relating to the heterogeneity between study estimates are given. The interested reader is again referred to the Cochrane online manual for further details. It is worth noting also that methods exist for identifying and correcting for, bias, and for assessing the confidence in the body of evidence.

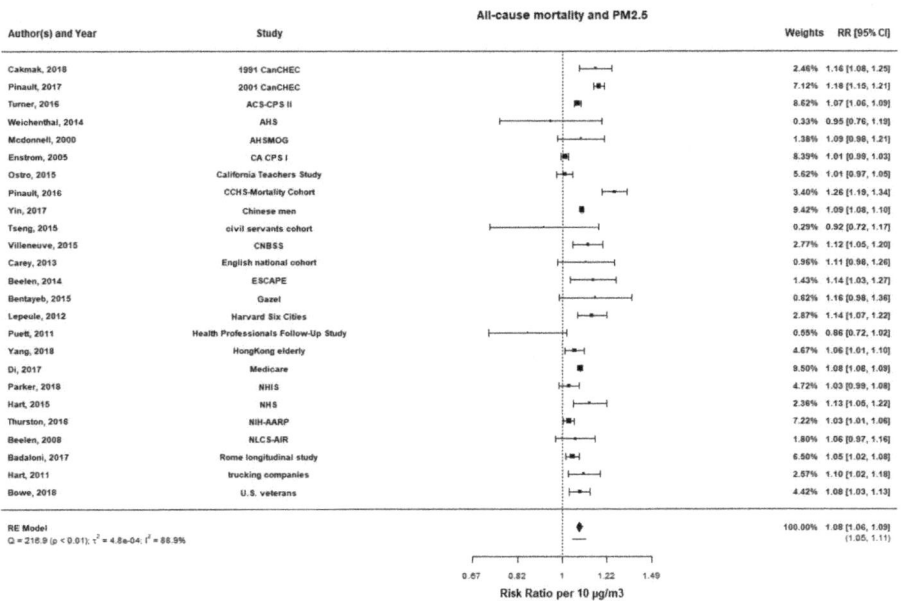

Figure 13.4 Forest plot for PM$_{2.5}$ and All-cause mortality.[13] Reproduced from ref. 13 with permission from Elsevier, Copyright 2020. Note: RR – relative risk; CI – confidence interval; Weights – percentage contribution to summary estimate; RE – random effects model; Q – Cochrane's Q statistic; p – *P*-value for Q; τ^2 – between study variance; I^2 – heterogeneity statistic.

References

1. Mortality and Morbidity during the London Fog of December 1952. Reports on Public Health and Medical Subjects No. 95. London: Ministry of Health; '50 years on' report.
2. BMJ Epidemiology for the uninitiated (bmj.com).
3. A. B. Hill, The environment and disease; association or causation?, *Proc. R. Soc. Med.*, 1965, **58**, 295–300.
4. K. J. Rothman and S. Greenland, Causation and causal inference in epidemiology, *Am. J. Public Health*, 2005, **95**(Suppl 1), S144–S150.
5. L. A. Cox Jr, Modernizing the Bradford Hill criteria for assessing causal relationships in observational data, *Crit. Rev. Toxicol.*, 2018, **48**(8), 682–712.
6. C. A. Pope 3rd, M. J. Thun and M. M. Namboodiri, *et al.*, Particulate air pollution as a predictor of mortality in a prospective study of U.S. adults, *Am. J. Respir. Crit. Care Med.*, 1995, **151**(3 pt 1), 669–674.
7. A. R. Khanolkar, R. Ljung and M. Talbäck, *et al.*, Socioeconomic position and the risk of brain tumour: a Swedish national population-based cohort study, *J. Epidemiol. Community Health*, 2016, **70**, 1222–1228.
8. Q. T. Ostrom, L. Bauchet, F. G. Davis, I. Deltour, J. L. Fisher, C. E. Langer, M. Pekmezci, J. A. Schwartzbaum, M. C. Turner, K. M. Walsh, M. R. Wrensch and J. S. Barnholtz-Sloan, The epidemiology of glioma in adults: a "state of the science" review, *Neuro-Oncology*, 2014, **16**(7), 896–913.

9. F. Forastiere, F. Lurmann and R. Atkinson, *et al.*, *Systematic Review and Meta-analysis of Selected Health Effects of Long-term Exposure to Traffic-related Air Pollution*, Health Effects Institute, Special Report, 23, 2022.

10. H. R. Anderson, A. Ponce de Leon, J. M. Bland, J. S. Bower and D. P. Strachan, Air pollution and daily mortality in London: 1987–1992, *BMJ*, 1996, **312**(7032), 665–669.

11. A. L. Cochrane, *Effectiveness and Efficiency: Random Reflections on Health Services*, Nuffield Provincial Hospitals Trust, London, 1972. (Reprinted in 1989 in association with the *BMJ*, Reprinted in 1999 for Nuffield Trust by the Royal Society of Medicine Press, London (ISBN 1-85315-394-X)).

12. A. L. Cochrane, 1931–1971: a critical review, with particular reference to the medical profession, in *Medicines for the Year 2000*, Office of Health Economics, London, 1979, pp. 1–11.

13. J. Chen and G. Hoek, Long-term exposure to PM and all-cause and cause-specific mortality: A systematic review and meta-analysis, *Environ. Int.*, 2020, **143**, 105974.

14 Causality

14.1 Introduction

This chapter differs from others in this book in that it contains no mathematics. The question of causality cannot be settled by the application of a formula or the solving of an equation, and the reader may wonder why we have included it. The reason is that the question has come to the forefront in work on air pollutants in recent years and, it seems to us, to follow on from our chapter on epidemiology. The fact that the question of causality cannot be settled by formulae or equations means that a range of opinions is often possible.

The problem of establishing a causal relationship between exposure to a potentially toxic compound and an effect on health is a difficult one. In this chapter we discuss the problem of deciding whether an association between some measure of air pollution and some index of ill-health is likely to be causal in nature. This problem is central to current discussions of the effects of air pollutants on health. We should be clear, from the outset, that it is impossible, or probably impossible, to prove that exposure to air pollutants is causally related to adverse effects on health. The difficulty of proving, *i.e.* being absolutely certain, that a *potentially* causal relationship *is* a causal relationship has troubled philosophers for as long as philosophy has been studied; we can hardly hope to resolve this difficulty here.

Causal relationships, or potentially causal relationships, are often discussed in terms of "cause and effect". We think the term "cause" is misleading; it implies that we know what causes the "effect". We also think the word "effect" is misleading in that it implies something has

Basic Mathematics for Students of Air Pollutants
By Robert Maynard and Richard Atkinson
© Robert Maynard and Richard Atkinson 2024
Published by the Royal Society of Chemistry, www.rsc.org

occurred as a result of the "cause", thus, assuming we know the solution to the very question we are trying to resolve. It would, we think, be clearer if we spoke in terms of associations between events. One event might be exposure to air pollutants; another admission to hospital. It is the relationship between these two events that interests us. If we can convince ourselves that they are causally related then we may speak in terms of "cause and effect"; if we cannot, then we must accept that the best we can say is that they are *associated*. We should define "associated". We take the word association to imply a statistically significant relationship; that is a relationship which seems unlikely to have arisen by chance.

Research into the effects of air pollutants on health is bedevilled by the problem of causality: event E is associated with pollutant P, but does P cause E? The problem is made more difficult by knowing that E is caused by factors other than P and that exposure to P, alone, never occurs. Air pollutants are invariably met as mixtures and separating the relationships between pollutants, between $P1$ and $P2$, and various events is difficult and, in some cases, probably impossible. Is this an important problem? The answer is yes, certainly if the regulation of concentrations of air pollutants follows a single pollutant approach with air quality standards being set for individual pollutants. Particulate matter presents a yet more difficult problem: it is invariably a complex mixture of chemical species.

In trying to decide whether an association between P and E, or more generally between A and B, is causal in nature, research workers have turned to Sir Austin Bradford Hill's paper: "The environment and disease: association or causation?".[1] This seminal paper was given, in London, as an address to the Section of Occupational Medicine at the Royal Society of Medicine and published in the Proceedings of the Society in 1965. That Bradford Hill's paper has been more admired than read is shown by frequent references to his work in terms of "criteria" for judging whether an association is causal in nature. Bradford Hill did not use the term "criteria" in his paper, nor did he speak of "tests" of causality. Bradford Hill was in fact careful to speak in terms of "features" of causal associations and argued that none of his now famous nine features of causal associations could be required before an association should be deemed to be causal or, perhaps better, likely to be causal. None, in his view, was an infallible test of causality. He argued that these nine features of causal associations, "help us to make up our minds on the fundamental question – is there any other way of explaining the set of facts before us, is there any other answer equally, or more, likely than cause and effect." He added that, "the decisive question is whether the frequency of the undesirable effect B will be influenced by a change in the environmental feature A".

A particularly insightful discussion of Bradford Hill's proposals has been contributed by Rothman and Greenland[2] although these authors headed their discussion "causal criteria" whilst acknowledging that Bradford Hill did not use the word "criteria". Rothman and Greenland argue that one of Bradford Hill's features of causal associations, temporality, is a requirement for acceptance of causality (if so, a true criterion) and that another, specificity of effect, is without merit.

Rothman and Greenland also provide an interesting discussion of the age-old problem of the value of induction as a method of arriving at truth and, in agreement with many philosophers, argue that it is an unreliable instrument. Popper's concept of falsification is discussed and accepted as a reliable means of deciding what is, at least, not true.[3] Falsification is widely accepted by research workers as an appropriate method: "bold conjectures", to use Popper's phrase, and rigorous attempts to falsify them have become the accepted hallmark of good quality science.

A detailed and critical appraisal of the Bradford Hill features of causal relationships has recently been published by Cox.[4] This review introduces a number of concepts and is recommended for those with a special interest in the assessment of causality. It is not possible, here, to give a summary of Cox's views but two points are worth mentioning, even briefly. First, Cox distinguishes sharply between deciding whether an observed association is causal in nature and deciding whether changing the putative cause will produce a change in the associated effect. He calls the latter "manipulative causation" and argues that a revision of the Bradford Hill "criteria" is required. It might be argued that air pollution scientists do not rely, only, on judgments of the likelihood of causality of observed associations to decide whether reductions in levels of air pollutants are likely to be associated with changes in indices of ill-health: they call upon other sorts of evidence including intervention studies; see below. Secondly, Cox introduces a number of methods which are not in current use in the air pollution field. Whether these will be adopted remains to be seen.

14.2   Features of Causal Associations

Bradford Hill identified nine features of associations likely to be causal in nature. The sort of associations of which he was thinking were those which had been identified by epidemiological studies and he began by defining the associations as unlikely to have arisen by chance or to reflect the activity of known confounding factors. This means that the associations with which he was concerned had already been shown to be statistically significant at conventional levels of confidence and that due account had been taken of confounding factors in so far as these were known before the question of causality was addressed. Associations that are not statistically significant were not discussed; it is usually accepted that a decision regarding statistical significance should be made before causality is considered. Failure to achieve statistical significance does not, of course, mean that the association is not a true association or indeed that it is not a causal association; it may simply mean that the study lacked the statistical power to determine whether the observed association was statistically significant. Testing for statistical significance involves calculation of a "p value" for the likelihood of the association having arisen by chance assuming the hypothesis of no association is true. For example if the critical value of p is set at 0.05 then p values of less than 0.05 will be deemed statistically significant. But we might have chosen 0.01 for the critical value; if the calculated value of p were 0.03 then a different view on statistical significance would be taken depending on whether the critical value was 0.05 or 0.01. Bradford Hill's features of causal associations will now be considered one by one.

14.2.1   Strength of Association

Bradford Hill suggested that strong associations were more likely to be causal in nature than weak associations. This has caused some confusion. There is no *a priori* reason to think that strong associations (a given change in A is associated with a large change in B) are more likely to be causal than weak associations (a given change in A is associated with a small change in B). Bradford Hill's point was that if a strong association is actually caused by a confounding factor that varies with A then that factor should not be too difficult to identify and if no such factor can be identified then this strengthens belief in causality. If on the other hand the association is weak then identifying the confounding factor or confounding factors might be very difficult. Use of the words "strong" and "weak" is perhaps unfortunate in that they seem to imply a little more than the size of the coefficient linking A with B. But finding better terms is not easy: almost any adjective might convey more than the magnitude of the coefficient.

Associations between ambient concentrations of air pollutants and potential effects on health are weak (the coefficients are small, the signal to noise ratio is low) and this has led some commentators to question whether they should be regarded as causal. Some have tried to identify a magnitude of an effect that should be exceeded before it should be regarded as large enough to be more or less immune from the effects of unknown confounding factors. A relative risk of 2 has been mentioned. There seems to be little basis for such a precise criterion: why not 1.5 or 3? Relative risks in the air pollution field are expressed "per unit increment in concentration" and, of course, if a larger unit increment is specified so will the relative risk also be increased. A relative risk of >2 has been taken by some to mean that B is "more likely than not" to have been caused by A and is used in that sense in some legal settings. Time-series studies, characteristically, have a small signal to noise ratio and report small relative risks and confidence intervals in some studies include zero. However, collections of these studies, when meta-analysed, show a positive association not likely to be explained by chance.

14.2.2 Consistency of Association

A widely accepted tenet of experimental science demands that before a finding can be accepted as true it must be replicated. This tenet or test has been a cornerstone of the interpretation of epidemiological evidence in the air pollution field. Cohort and time-series studies conducted all over the world, with different investigators and different data, have reported similar findings that support causality. The use of consistency as a feature of causal associations also reflects a belief that if a result can be replicated in several different laboratories by different investigators using different methods it will be unlikely to have arisen as a result of conditions obtained in one specific laboratory. This faith in multiplying the tests is open to argument; a critical factor leading to a false conclusion might exist in all the laboratories. An example is provided by the use of a single strain of laboratory rat. These rats are highly in-bred and one rat is genetically very similar to the next. Repetition undertaken with a view to exploring inter-subject variability or to reduce the effects of some laboratory-specific factor will be of limited value if such a single strain is used. An experienced toxicologist pointed out to one of the present authors: "If you test a compound on a hundred highly inbred rats what you have done is to test the compound on one rat one hundred times." Exaggeration? Yes, no doubt, but not entirely so.

Experimentalists tend to distrust absolutely consistent results: absolute consistency is rare in nature and its occurrence suggests to the cautious worker that some unrecognised factor is in play. If all the time-series studies of the effects of ambient particles on health, undertaken in different parts of the world, had produced exactly the same results one might have concluded that something odd was going on. One implication of such a set of results would have been that there was no variation in effect-modifying factors from place to place. Another is that the composition of particulate material, known to vary from place to place, is unimportant in causing the reported effects. This seems unlikely. A reasonable consistency between studies is certainly encouraging; a modest amount of study to study variation in results is also reassuring. Another of the tenets of science is that heterogeneity provides clues as to how nature works. Reasons for variations in results from one study to another should always be sought.

14.2.3   Specificity of Effect

Bradford Hill argued that an association between some factor and a single effect was, with reservations, more likely to be causal than an association between a factor and a number of effects. Rothman and Greenland described this "criterion" as "useless and misleading". Bradford Hill may have been encouraged in his belief by the advances in bacteriology during his long lifetime. He would have known of the work of Koch and the precepts known as Koch's Postulates.[5] These were a series of what might be regarded as rules put forward by Koch for ascertaining whether a bacterium was the cause of a given disease.

1. The microorganism must be found in large numbers in all organisms suffering from the disease, but should not be found in healthy organisms.
2. The microorganism must be isolated from a diseased organism and grown in pure culture
3. The cultured microorganism should cause the disease when introduced into a healthy organism.
4. It must be possible to isolate the microorganism from the inoculated, diseased experimental host and it must be identified as being identical to the original specific causative agent.

It will be seen that these represent rather more prescriptive views than those expressed by Bradford Hill. Here the emphasis is on specificity of effect: one bacterium – one disease. Rather similar "rules" were put forward by Dale for use in deciding whether a chemical compound was acting as a transmitter substance at nerve terminals.[6] Again the emphasis was on specificity. At the time Bradford Hill was writing it was accepted that cigarette smoking was likely to be a major cause of lung cancer but that it was also a cause of manifold other diseases was not. That he regarded specificity of effect as a feature of causal associations is not, in retrospect, surprising. If there is specificity of effect, as perhaps in the association between exposure to asbestos and the occurrence of asbestosis, then the case for causality is strengthened. But a lack of specificity does not fatally weaken the case for causality.

Studies of air pollutants and health have shown that ambient particulate matter is associated with a wide range of health endpoints including deaths from respiratory and cardiovascular disease, exacerbations of respiratory and cardiovascular diseases, and, rather more surprisingly, dementia, birth defects and low birth weight. At first glance this range of potential effects is worrying with regard to the likelihood of causality: can air-borne particles really produce such a wide range of conditions in such a wide range of locations within the body? The answer may well be yes if it can be shown that particles exert their effects by a mechanism which can play a part in causing all these effects. Perhaps this is the key: specificity of mechanism rather than specificity of effect in terms of health endpoints, strengthens the case for causal relationships.

14.2.4 Temporality

Most commentators agree that this is a requirement of causal associations. Bradford Hill, rather oddly perhaps, did not regard it as such and pointed out that though in principle "the horse should precede the cart", in practice distinguishing the horse from the cart might not be easy. He pointed out that this might be especially the case with regard to "diseases of slow development" and asked, "Does a particular diet lead to a disease or do the early stages of the disease lead to those peculiar dietary habits?" The case for temporality is most easily made when the putative causal agent appears at an identifiable point in time. A appeared on Tuesday and B appeared on Wednesday, or A was reduced on Thursday and B declined on Friday. Such occurrences are rare: in the air pollution field, exposure varies from day to day and from year to year, but everybody is exposed to some level of air pollution for all of their lives. Cardiovascular disease is also thought to develop over long periods. Temporality may sometimes be difficult to confirm but natural experiments and deliberate interventions allow it to be identified with some confidence.

14.2.5 Biological Gradient

This feature of causal associations suggests that as A is increased B should increase and that as A is reduced B should decline. Most toxicologists would recognize this as an encouraging finding. It applies to graded responses to increasing doses of drugs and to the number of cells, animals or people showing a specified response as the dose is increased. For example, the number of people dying per day tracks with the daily average concentration of particles measured as PM_{10}. Of course, the relationship might be misleading: if a confounding factor, C, which can cause B, varies with A; then B will also vary with A.

14.2.6 Biological Plausibility

Neither Bradford Hill nor Rothman and Greenland regarded this feature as critical: associations which seem inexplicable today may be explained tomorrow. No doubt this is true but many people, including some scientists, find it easier to accept something which can be explained than something which cannot. Bernard's remark, "... it is what we think we know that prevents us from learning"[7] should be recalled when one is tempted to discard the suggestion that an association is causal on the grounds that it cannot be explained. Evidence from toxicological studies provides some support for the assertion that associations between low concentrations of air pollutants and effects on health are causal in nature but this evidence has not convinced all commentators. In recent years a new sort of support has been provided: the support of the plausible hypothesis. Seaton et al.[8] suggested that the associations between low mass concentrations of particulate matter and effects on the cardiovascular system might be explained by the effects of the large number of ultrafine particles found in ambient air. This exciting hypothesis has been remarkably fruitful in generating ideas for experiments and has not yet been falsified; indeed, much evidence supports it. Has the existence of this hypothesis strengthened the case for regarding the associations between ambient concentrations of particles and effects on health as causal in nature? We think the answer is yes, and probably a little more than it should have done. That a plausible explanatory hypothesis can be conceived suggests that the association may well not be inexplicable; but we should not confuse hypotheses and evidence.

14.2.7 Coherence

Bradford Hill suggested that the cause and effect interpretation of an association between a putative cause and a disease should not "seriously conflict with the generally known facts of the natural history of the disease". He pointed to the coherence between the histopathological evidence from the bronchial epithelium of smokers and the isolation from cigarette smoke of factors shown to be carcinogenic when applied to the skin of laboratory animals. This seems to be more an argument for biological plausibility than for coherence. In the air pollution field coherence has been seen as being demonstrated by associations between ambient concentrations of air pollutants and, for example, deaths from cardiovascular disease, admissions to hospital for the treatment of cardiovascular diseases and changes in levels of clotting factors in the blood. These effects are coherent: they fit together and add to the likelihood that the associations are causal. Bates[9] pointed out that this sort of coherence occurred in the air pollution field: his observation has been accepted as supporting the case for causality.

14.2.8 Experimental Evidence

Natural experiments, experiments that occur not as a result of deliberate human activity, occur all the time in the air pollution field. For example, concentrations of air pollutants vary from day to day as a result of changing weather conditions. Episodes of unusually high ambient concentrations occur less often but the reader will know that air pollution episodes such as the London smog of 1952 stimulated interest in our field. In addition to such natural experiments air pollution research workers have taken advantage of policy decisions which have led to reductions in ambient concentrations of air pollutants, for example deliberate reductions in traffic density during the Olympic Games held in Beijing in 2008. These studies are known as intervention studies: the effect of an intervention is studied. Two major reviews of intervention studies have been published recently: Burns[10] reported the findings of an "abridged Cochrane systematic review" of intervention studies; Henschel[11] reported the findings of a less systematic review. The latter study found consistent evidence to show that decreases in ambient concentrations of air pollutants resulted in benefits to health. The former review found the evidence less persuasive: it was clear that the evidence was mixed (some studies reported benefits, some did not), but none reported harmful effects of reduced ambient concentrations. In addition, there is the mass of experimental work undertaken with human volunteers and laboratory animals. Our field does not lack experimental evidence of the effects of air pollutants on health.

14.2.9   Analogy

Bradford Hill was brief in dealing with this feature; he pointed to the effects of thalidomide and of rubella on the foetus and argued that these effects made it easier to believe that associations between exposure to other drugs or viral infections and impaired development might be causal. Rothman and Greenland[2] seemed rather unimpressed by this feature and argued that it could be used as a source of hypotheses but that failure to think of an analogy could not be regarded as falsification of the hypothesis that an association was causal. The former point is obvious; the latter point is undoubtedly true. Toxicologists often argue by analogy though they tend to use the term "structure–activity relationships" when arguing that a chemical is likely to have effects similar to those of chemicals to which it is closely related. Such arguments have been shown to be valid for many groups of compounds.

14.3   How Sure Should We Be About Causality Before Recommending Action?

The reader will by now be aware that certainty about whether associations are or are not causal in nature is likely to evade us. Bradford Hill knew this and carefully described his "features" of causal relationships as features and not as criteria. Cox[4] asked an important question in his analysis of the problem: does associative causality imply manipulative causality? By "associative causality" Cox means, unless we have misread his argument, the sort of evidence of associations between air pollutants and events indicative of ill-health which is provided by most types of epidemiological study: that B is likely to be causally linked with A. By "manipulative causality" he meant, again unless we have misread his argument, demonstration that reductions in A are likely to be followed by reductions in B. Many have assumed that this is the case: the observation that when water is heated the mercury in the thermometer rises suggests, to most people very strongly, that when the water cools the mercury will fall. Cox argued that associative causality does not provide an acceptable level of certainty. His arguments on this point are cogent and would, we think, be accepted by many epidemiologists working in the air pollution field. However, we also think that Cox, by focusing on associative causality as a basis for assuming manipulative causality, has ignored other sorts of evidence that bear on the question.

In our view, decisions of whether reductions in levels of air pollutants are likely to be followed by reductions in events indicative of ill-health should be taken after review of all the available evidence. This includes our knowledge of the effects of episodes of high concentrations of air pollutants, intervention studies, and experimental studies as well, of course, as epidemiological studies that report associations between ambient levels of air pollutants and events indicative of ill-health. This mass of evidence cannot, at least not currently, be fed into a computer with the expectation that an irrefutable answer will be produced. Certainty will elude us.

In the absence of a means of reaching certainty we must rely on our judgment: a point stressed by Bradford Hill. To make an analogy: no single brick in the wall can support the weight of the roof, we must judge whether the collection of bricks that form the wall can do so. Perhaps the most important question to consider when thinking about whether an association is likely or not to be causal is: can we think of a better explanation for the association? If the answer is no then it would be sensible to accept the explanation of causality until a better explanation appears. In the sort of work we are considering it is possible that a better explanation will appear. Having decided whether the associations reported by epidemiologists are likely to represent casual associations we must ask a second question. This is: given that the associations are causal in nature, do they suggest that reducing levels of air pollutants will be of benefit to health? Again, all the evidence must be called for examination. If the answer is yes, then a third question: how much benefit is likely to be delivered by a specified reduction in levels of air pollutants? Here we have fewer guides than is the case regarding the previous questions. At present, meta-analysis of the findings of epidemiological studies seems to be our best guide.

14.4   Scientific Caution and the Precautionary Principle

In any discussion of the effects of air pollutants on health and what should be done, sooner or later somebody will mention the precautionary principle. This principle encourages policy makers *not* to wait for complete certainty before they act to reduce some threat to health. It is equally certain that somebody will mention cigarette smoking and point out that had policy makers acted sooner many lives would have been saved.

It is important to remember that there is a difference between scientific caution and the precautionary principle. All scientists are taught to be critical of evidence put before them; they are taught to be cautious about accepting evidence. The search for flaws must be rigorous. This process is the first step in assessing evidence of the effects of air pollutants on health with a view to deciding whether or not action to reduce levels of air pollutants should be taken. It should be obvious that as complete as possible a collection of evidence should be undertaken. The US Clean Air Act requires the US Environmental Protection Agency (EPA) to undertake such reviews at specified intervals. In many countries the guidance set out by the Cochrane Collaborations is used as instructions for how evidence should be collected and reviewed. Meta-analysis often follows the collection of evidence; the desire to summarise the evidence as a coefficient qualified by confidence intervals is strong, perhaps a little too strong. Meta-analysis, of course, provides its own problems: publication bias and how to handle missing data provide examples. Methods for dealing with such problems have been developed and are well explained by Rothman and Greenland.[2] Once all the evidence, or as much as can be collected, has been collated, reviewed and summarised, perhaps by meta-analysis, it must be weighed or graded. In our field the question is likely to be, do we accept that exposure to ambient concentrations of some air pollutant or combination of air pollutants is causing an effect or effects, on health? The word "accept" is slightly vague: it might mean accept without reservations, accept with reservations, accept provisionally and so on. Returning to our analogy of bricks holding up a roof we might say that the more evidence there is, the better, in terms of quality, the studies that have produced the evidence; the more varied the evidence, the more closely the evidence from different types of study fits together, then the more firmly we should accept the proposition. Several attempts have been made to define confidence in such assertions: the evidence may be graded in terms of its nature, of its quality and relevance. Any such grading system involves the application of judgment: experts may differ as to the weight they attach to various strands of evidence. Consensual or majority decisions made by groups of experts are required. All this process should involve scientific caution and critical thinking.

The findings of expert assessment of the evidence are presented to policy makers: the representatives of, or those working for the representatives of, the public. If the advice from the expert review is clear cut in the sense that the experts advise that the case for a causal association is strong or weak, then the policy maker's task is made easier: he or she may, or may not, proceed to the next step which involves weighing the likely costs and benefits of policy options. However, it is when the advice from experts is not clear cut that the policy maker needs another tool to help in making the decision. This tool is provided by the precautionary principle. People may vary in their definitions of the precautionary principle. Here, we shall note the explanation of the principle provided by the Commission of the European Communities in 2000 (Brussels, 02.02.2000. COM(2000) 1). This communication is worth looking at in detail; readers are advised to do so. It will be seen that the precautionary principle is not carte blanche for acting in the face of all doubt; indeed, it is far from that.

The problem of whether associations between concentrations of air pollutants and events indicative of ill-health are causal in nature is one of the central problems of air pollution science. The problem has come to the fore in recent years when associations have been reported at concentrations which are far below those of the past. At the time of the London smog of 1952[12] there was little doubt that high concentrations of air pollutants had effects on health. That such incidents were undesirable was patent; that action should be taken was obvious. Modern epidemiological studies have shown that effects may occur at low, in some cases very low, ambient concentrations. As the size of the effects on health has fallen, doubt about the continuing causality of the asociations has risen: deciding whether associations represent causal relationships has become more diffcult. Careful application of Bradford Hill's advice about the features of causal associations remains a central part of the decision making process.

References

1. B. Hill and S. Austin, The environment and disease: association or causation?, *Proc. R. Soc. Med.*, 1965, **58**, 295.
2. K. J. Rothman and S. Greenland, *Modern Epidemiology*, Lippincott, Williams and Wilkins, Philadelphia, 2nd edn, 1998
3. K. Popper, *Conjectures and Refutations*, Routledge and Kegan Paul, London, 3rd edn (revised), 1969.
4. L. A. Cox Jr, Modernizing the Bradford Hill criteria for assessing causal relationships in observational data, *Crit. Rev. Toxicol.*, 2018, **48**(8), 682.
5. R. Koch, *Ueber bakteriologische Forschung. Verhandlungen X internat. Med. Kongr*, Hirschwald, Berlin, 1890, 1. (Information taken from: Chapter 43, The Mechanisms of Bacterial Infection, in *Topley and Wilson's Principles of Bacteriology and Immunity*, ed. G. S. Wilson and A. M. Sri, Edward Arnold, London, 5th edn, 1964).

6. Dale, Sir Henry, Though a set of criteria for proving that a putative neurotransmitter is a neuro-transmitter are almost invariably attributed to Dale it is curiously difficult to find any evidence that Dale set out such a list. The criteria should not be conflated with Dale's law which says that a single neuron produces only one transmitter substance: now known to be incorrect. For a list of criteria of the sort usually attributed to Dale see: W. C. Bowman and M. J. Rand, *Textbook of Pharmacology*, 2nd edn, Blackwell Scientific Publications, 1980, p. 9.18: criteria for identifying a chemical transmitter. Dale's review: Dale, Sir Henry: Walter Ernest Dixon Memorial Lecture 1934. Pharmacology and nerve-endings, *Proc. R. S. Med.*, 1935, 28, 15 should be consulted for further information.

7. Claude Bernard: quotation taken from: *Hannibal's March* by Sir Gavin de Beer, Sidgwick and Jackson, London, 1967.

8. A. Seaton, W. MacNee, K. Donaldson and D. Godden, Particulate air pollution and acute health effects, *Lancet*, 1995, **354**, 176.

9. D. V. Bates, Health indices of the adverse effects of air pollution: The question of coherence, *Environ. Res.*, 1992, **59**, 336.

10. J. Burns, H. Boogaard and S. Polus, *et al.*, Interventions to reduce ambient air pollution and their effects on health: an abridged Cochrane systematic review, *Environ. Int.*, 2020, **135**, 105400.

11. S. Henschel, R. Atkinson, A. Zeka and A. Le Tertre, *et al.*, Air pollution interventions and their impact on public health, *Int. J. Public Health*, 2012, **57**, 757.

12. Ministry of Health: Reports on Public Health and Medical Subjects, Number 95. Mortality and Morbidity during the London fog of December 1952, HMSO, London, 1954.

15 Quantification of the Effects of Air Pollutants on Health

15.1 Introduction

It is widely accepted that short-term exposure to current ambient concentrations of air pollutants has effects on health. These effects range from deaths to hospital admissions, to consultations with general practitioners, and to the occurrence of symptoms relating to the respiratory and cardiovascular systems. In addition, long-term exposure to particulate matter monitored as $PM_{2.5}$ has been shown to cause increased risks of death from cardiovascular disease and from lung cancer. It is obviously important to estimate how large these effects actually are. One would wish to know this on the grounds of scientific curiosity but, and more important, one also needs to know before one can undertake calculations of the benefits to public health likely to accrue from reductions in concentrations of air pollutants. These estimates feed into cost–benefit calculations used to justify, in part, possible levels of public spending on reductions of levels of air pollutants. These cost–benefit calculations also have to be weighed against those for competing risk factors. We discussed, in Chapter 13, the elements of the epidemiological techniques used to study the effects of air pollutants on health and noted that methods reveal associations between ambient concentrations of air pollutants and measures of effects on health. In Chapter 14 we discussed the examination of such associations with a view to deciding whether or not they were likely to be causal in nature. In this chapter we assume that associations

Basic Mathematics for Students of Air Pollutants
By Robert Maynard and Richard Atkinson
© Robert Maynard and Richard Atkinson 2024
Published by the Royal Society of Chemistry, www.rsc.org

have been reported and, on examination, considered to be likely to be causal, and proceed to use the mathematical measures of associations, the coefficients linking concentrations with effects, to estimate the impact of air pollutants on health. We have focused on estimation of estimates of deaths attributable to exposure to air pollutants and on ways of reflecting this impact on public health.

Let us begin with the effects of long-term exposure to fine particles, monitored as $PM_{2.5}$, on the risk of death. By "death" we mean death from all causes excluding accidents. To make the calculation we shall require a coefficient linking long-term ambient concentration with the risk of death. The American Cancer Society cohort studies[1,2] have shown that long-term exposure to a concentration of fine particles ($PM_{2.5}$) of 10 µg m^{-3} is associated with a relative risk of death (all causes, non-accidental, mortality) of 1.06 (95% CI: 1.02–1.11). We shall ignore the confidence intervals for the moment and note, only, that they imply a conventionally significant statistical relationship.

Approximately 1% of the UK population dies each year from non-accidental causes. Thus, there are about 500 000 deaths per year. We may calculate the fraction of these deaths attributable to long-term exposure to fine particles.

15.2 Application of Relative Risk to Calculation of Effects of Air Pollutants

Let us consider two equal populations: A and B. Let us assume that A and B are both exposed, over a long period to air pollutants but that the average exposure of population B is greater than that of population A.

For a given time period let risk of death in population A = r (expressed as number of deaths per 1000).

Let relative risk of death in population B compared with population A = RR.

Then the risk of death in population B in given time period = RR × r (expressed as deaths per 1000 of the population).

If we denote the number of people in population B as N, then:
number of deaths in population B = RR × r × N.

We can then calculate the "excess" or extra deaths in population B compared with population A = RR × r × $N - r$ × N.

Excess deaths as fraction of deaths in population B = (RR × r × $N - r$ × N)/RR × r × N

= r × N(RR − 1)/RR × r × N

= (RR − 1)/RR.

So, by knowing only the RR we can calculate the excess number of deaths in a more-exposed population (B) relative to an less-exposed population (A). In the context of air pollution, we define the more-exposed and less-exposed populations by the average pollution concentrations to which they are exposed and calculate the excess deaths associated with the differences in these exposures. This is determined by the specification of the RR increment. For example, in the ACS study quoted above, the RR for $PM_{2.5}$ was 1.06 for a 10 µg m^{-3} increment in $PM_{2.5}$. For a smaller increment, then the RR will be smaller. It is also worth noting that the increment is not dependent upon the pollutant concentrations. The difference could be between 5 and 10 µg m^{-3} or 15 and 25 µg m^{-3}, *etc.*

Let us illustrate this with an example.

Let the RR associated with long-term exposure to an increase of 10 µg m^{-3} $PM_{2.5}$ be 1.06.

Excess deaths as a fraction of all deaths in population exposed to 10 µg m^{-3} $PM_{2.5}$ more than another population = $(1.06 - 1)/1.06 = 0.0566$.

Let annual number of deaths in UK be 500 000. This is the number of deaths per year in the exposed population.

Number of deaths attributable to long-term exposure to $PM_{2.5}$ = $0.0566 \times 500\,000$

$= 28\,300.$

Let, without any loss of generality, the annual average $PM_{2.5}$ = 12 µg m^{-3}.

This simple calculation yields an impressive result: about 28 000 deaths per year are attributable to long-term exposure to fine particles ($PM_{2.5}$) at the current ambient concentration of 12 µg m^{-3} *compared* with a population exposed to $PM_{2.5}$ at an ambient concentration of 2 µg m^{-3} (*or counterfactual*). Such a figure appears, frequently, in media articles that draw attention to the effects of air pollution on health.

An even simpler, and slightly less accurate, calculation may be made. If the total number of deaths is 500 000 and this is 6% more than it would have been had there been long-term exposure to fine particles ($PM_{2.5}$) of 2 µg m^3; then the excess deaths attributable to long-term exposure to a concentration of fine particles ($PM_{2.5}$) of 12 µg m^{-3} may be estimated: if x is the number of deaths without exposure to $PM_{2.5}$ then an extra 6% attributable to exposure equals $1.06 \times x = 500\,000$ deaths per year. Hence $x = 471\,689$ and 28 311 of the 500 000 deaths per year are attributable to $PM_{2.5}$ exposure.

An even less accurate estimate may be made by calculating 6% of $500\,000 = 30\,000$.

What do these numbers mean? They mean that if the annual average concentration of fine particles ($PM_{2.5}$) continued at 12 rather than

$2~\mu\mathrm{g~m}^{-3}$ we could make this calculation every year and arrive at the same result. Each year we could say that there was an excess of about 28 000 deaths as a result of long-term exposure to fine particles. On a world scale the numbers are, of course, much larger. In a paper published in 2017[3] it was estimated that "Ambient $PM_{2.5}$ was the fifth-ranking mortality risk factor in 2015. Exposure to $PM_{2.5}$ caused 4.2 million (95% uncertainty interval [UI] 3.7 million to 4.8 million) deaths...". The authors added that 59% of these deaths occurred in East and South Asia.

One interpretation is that the figure of 28 000 (deaths associated with $PM_{2.5}$) can be compared with a UK figure of 1790 for deaths from road accidents or about 40 000 deaths from lung cancer or 12 000 deaths from prostate cancer.

Another interpretation is that the figure of 28 000 deaths associated with $PM_{2.5}$ is misleading in that it suggests that all remaining deaths (let us say 500 000 minus 28 000 = 472 000) had nothing at all to do with exposure to particulate matter monitored as $PM_{2.5}$. If this were true then we should have to ask, what was it about the 28 000 people that made them sensitive to $PM_{2.5}$? It must have been something rather remarkable because the other 472 000 people were not sensitive at all. This seems, to us, to be unlikely.

It seems more likely to us that all people are, to a greater or lesser extent, affected by long-term exposure to $PM_{2.5}$. What we actually know is that the risk of death is increased *on average* by 6%. For some people, for example those with heart disease, the increase in risk might well be greater than 6%, for others the increase in risk might be less than 6%. By using the average we lose this perception of variability of increase in risk. It would be correct to say that the effect *is equivalent to* 28 000 deaths per year. This is, in fact, what was said in the report of the UK Committee on the Medical Effects of Air Pollutants.[4] Another term used to describe the results of this sort of calculation is "attributable deaths". In the work done by COMEAP the number of attributable deaths was estimated to be 29 000 and we shall use this number, for illustrative purposes only, from now on.

15.3   The Use of Life Tables

Another, and in our view much more meaningful, approach to representing the effect has been developed.[4] This method involves the use of life tables and is, at first glance, complicated. This is not actually so and the method is easily explained. Before we start it is worth clarifying an important concept. First, the risk of death for everyone is 1 (or 100%) – everyone dies eventually. When we talk about a pollutant increasing the risk death we are really considering survival – if you are exposed to a pollutant does it shorten your life compared to not being exposed? This concept is embedded in the analysis of cohort studies (see Chapter 13) that estimate survival in populations exposed to different concentrations of pollutants. Cohort studies provide the evidence for assessing the impact of air pollution on health. The next questions is then by how much does exposure shorten your life? A day or two may be inconsequential, but a number of years is clearly not. Let's now consider life tables and how they help us understand the impact of long-term exposure to air pollution on human health.

A life table provides figures for the statistical risk of death at all ages. The risk is, for example, low in the 20–30 age group and comparatively high in the 70–80 age group. Current risks, of course, reflect the effect of exposure to air pollutants. If we want to find out the effect of reducing the annual average concentration of fine particulate monitored as $PM_{2.5}$ by, for example, 1 µg m^{-3}, all we have to do is to reduce the risk, for each age group (actually each adult age group because we know, only, about effects of air pollutants on the risk of death in adults), by 0.6%. We recall that a 10 µg m^{-3} increment was associated with a 6% increase in risk of death. We then take a hypothetical cohort of, for example, all the people alive in 2019 and allow them to work their way, so to speak, through the table. In each year some will die. Those who don't will progress, so to speak, to the next year and to the next risk of death. If we follow the progress of the cohort for, let us say, 100 or so years then the entire cohort will have died. We should note that all members of the cohort will, eventually die. This means that the question, how many lives are saved by reducing $PM_{2.5}$, is meaningless. The much more important question which we must now ask is, how many years will the cohort have lived? What we mean is how many years of life, or life-years, will the cohort have enjoyed. And, very importantly, the total will have *increased* as a result of the reduction in $PM_{2.5}$. Putting the question in another way: what is the increase in average life expectancy?

The calculation could be made more complicated but also more realistic, by adding new births every year to the cohort. These calculations would be unbearably tedious if we had to do them by hand but using a computer makes them feasible. The results of the calculation undertaken by COMEAP are shown in Table 15.1. The only figure from Table 15.1 that we shall focus on here is the figure of 4 084 000 life years gained for a 1 µg m^{-3} reduction in $PM_{2.5}$.

Table 15.1   Life-years gained (rounded to the nearest 1000) over 106 years, by population, including new births, following specified a 1 μg m^{-3} reduction in PM$_{2.5}$. UK totals are aggregated from the individual results presented. Reproduced from ref. 4, https://assets.publishing.service.gov.uk/government/uploads/system/uploads/attachment_data/file/304641/COMEAP_mortality_effects_of_long_term_exposure.pdf, under the terms of the Open Government Licence v3.0, https://www.nationalarchives.gov.uk/doc/open-government-licence/version/3/.

Country	Life-years gained	Life-years gained per 100 000 people aged 30 years and over
England and Wales	3 604 000	10 597
Scotland	353 000	10 687
Northern Ireland	128 000	12 302
UK total	4 084 000	10 651

Table 15.2   Increased life expectancy for UK population following reduction of 1 μg m^{-3} in PM$_{2.5}$. Reproduced from ref. 4, https://assets.publishing.service.gov.uk/government/uploads/system/uploads/attachment_data/file/304641/COMEAP_mortality_effects_of_long_term_exposure.pdf, under the terms of the Open Government Licence v3.0, https://www.nationalarchives.gov.uk/doc/open-government-licence/version/3/.

Country	Increased life expectancy (days) for the 2008 birth cohort	
	Males	Females
England and Wales	21	20
Scotland	23	21
Northern Ireland	22	21
UK[a]	21	20

[a]Calculated by weighting the England and Wales, Scotland and Northern Ireland results by relevant birth cohort size.

The calculation also yields data for the increase in life expectancy of the population. A selection of figures from the COMEAP report is shown in Table 15.2. It will be seen that for a 1 μg m^{-3} reduction in PM$_{2.5}$ the average (across 2008 birth cohort) increase in life expectancy is about 20 days.

Another very interesting set of results is shown in Table 15.3, again from the COMEAP report. Here we are looking at estimated loss of life expectancy caused in 2008 by long-term exposure to fine particles. If we look at all deaths (over the age of 30 because we know only about effects on adults) the loss of life expectancy is about 6 months. This is the *average* effect across all deaths. If we look at the last line of

Table 15.3   Hypothetical average years of life expectancy lost in 2008 due to the contribution of anthropogenic particulate air pollution, averaged over different sections of the UK population. Data from ref. 4

Hypothetical population affected	Number affected	Hypothetical average loss of life expectancy
Whole population (ages 30+)	38 348 000	3 days
All deaths (ages 30+)	569 000	0.5 year
50% of deaths (30+)	290 000	1 year
Deaths from CV[a] causes (30+)	191 000	2 years
20% of deaths (30+)	116 000	3 years
10% deaths (30+)	58 000	6 years
7% deaths (30+)	40 000	8.5 years
"Attributable" deaths (30+)	29 000	11.5 years

[a]Cardiovascular.

the table we see the figure of 29 000 deaths appears. If these deaths, alone, were affected then the average loss of life expectancy would be more than 11 years. These figures, 6 months and 11 years, give us the boundaries of the effect: not less than about 6 months; not more than about 11 years.

It will by now be clear that there are several ways of expressing the effects of air pollutants on health.

We have mentioned the COMEAP report[4] several times. Let us now examine the conclusions of the report. We have condensed the statements made by the original authors.

Removing all anthropogenic ("human made") particulate matter air pollution (measured as $PM_{2.5}$) could save the UK population approximately 36.5 million life years over the next 100 years and would be associated with an increase in UK life expectancy from birth, i.e. on average across new births, of six months. This shows the public health importance of taking measures to reduce air pollution.

A policy which aimed to reduce the annual average concentration of $PM_{2.5}$ by 1 $\mu g\ m^{-3}$ would result in a saving of approximately 4 million life years or an increase in life expectancy of 20 days in people born in 2008.

The current (2008) burden of anthropogenic particulate matter air pollution is, with some simplifying assumptions, an effect on mortality in 2008 equivalent to nearly 29 000 deaths in the UK at typical ages and an associated loss to total population life of 340 000 life-years. This burden can be represented as a loss of expectancy from birth of approximately six months.

The uncertainties in these estimates need to be recognised: they could vary from about a sixth to double the figures shown.

The use of life table calculations leads to some interesting findings regarding the effects of reducing concentrations of air pollutants. We have already noted that in the long run no lives are "saved". Of course not, everybody dies at some time. Let us examine Figure 15.1, adapted from the COMEAP report to which reference has already been made. Let us look, first, at the lower panel (c) of Figure 15.1. Let us focus on the dashed line and, for the moment, ignore the solid line. The graph shows the effects on annual mortality (deaths per year) of a cohort of people (all the people alive in 2008) of making a sudden reduction of 1 μg m^{-3} in the ambient concentration of fine particles (PM$_{2.5}$). Note that in the first year after the reduction in PM$_{2.5}$ there is a sharp reduction in annual deaths: the graph shows that they are reduced by about 2900 (of course! A tenth of 29 000 because we are dealing with 1 and not 10 μg m^{-3}). But, as we follow the graph, we see that the reduction in annual mortality falls year by year until by 2050 it has returned to the level at which it stood in 2008. Even more remarkably, at least at first sight, the reduction in annual mortality then changes sign and become negative, meaning an increase in annual mortality. What is going on? The answer requires a little thought.

What has happened at the start of the period is that some of those who would have died in year 1 have survived to year 2. They are now a year older, and their likelihood of dying is greater than it was in year 1. This offsets the reduction in deaths caused by the reduction in PM$_{2.5}$. The process is repeated again and again: the cohort is aging and more people are living longer. Because everybody dies eventually this gain in length of life lived has to be "paid for": by 2050 the cohort is a good deal older (meaning that people have lived longer than they would have done had PM$_{2.5}$ not been reduced) and the annual death rate begins to rise. It goes on rising and the cohort diminishes in size. As it diminishes the number of deaths per year falls again and by 2113 all the cohort has died. The area labelled A is identical with the area labelled B.

This is very important. If we think in terms of a cohort (in our example, all the people alive in 2008) then in the long run no lives are "saved". But, and this is a very big "but", the members of the cohort live for longer than they would have lived had the reduction in PM$_{2.5}$ not been made.

If we now look at the first panel (a) of Figure 15.1 we can see this effect of the cohort living longer in sharp focus.

The dashed line shows the number of life years gained per year by the cohort as we proceed through the period 2008 to 2113. Note that in 2050 an additional 42 000 life years is lived by the cohort; this

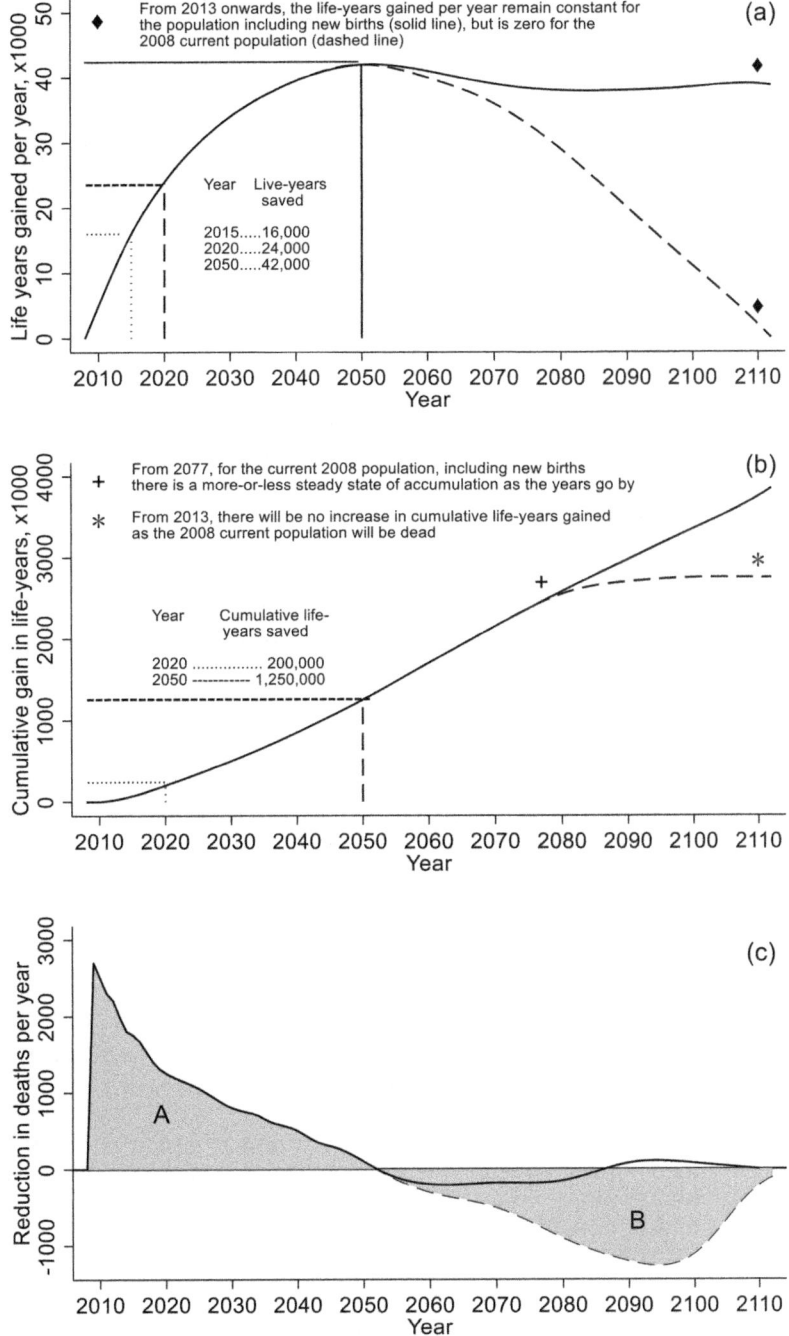

Figure 15.1 Patterns of impacts following a permanent reduction of 1 μg m^{-3} in annual average PM$_{2.5}$ concentrations impacting on all-cause mortality hazard rates for England and Wales. Impacts expressed

(continued)

is the peak annual gain in years of life lived and it means that a lot more people would be alive in 2050 than would have been the case had $PM_{2.5}$ not been reduced. Note, then, that after 2050 the number of additional life years lived per year by the cohort falls year by year until all the members of the cohort have expired, by 2113. The total life years saved or, if you prefer, the total additional years of life lived, is represented by the area under the curve. This gain in years of life lived by the cohort is the real, very significant, effect of reducing $PM_{2.5}$, in 2008, by $1\ \mu g\ m^{-3}$. It is a large effect: a large benefit.

Panel (b) of Figure 15.1 shows the effect in terms of the cumulative number of additional years lived by the cohort as we move through the period 2008 to 2113. Note again the dashed line: it rises and then flattens. By 2113 it is horizontal; this means that all the additional years of life lived have now been gained. All members of the cohort have died and, obviously, no more years of life can be accumulated. Now look at the scale on the y axis: it shows the cumulative gain in years of life lived (life-years) × 1000. By 2050 more than 1 000 000 years of life have been added; by 2113 the number is approaching 3 000 000. Nobody could deny that these are major benefits.

If you have grasped the points made above, you will have no trouble in understanding the next paragraph. Before reading on make sure that you do grasp the points made above.

So far, we have considered the effects of a $1\ \mu g\ m^{-3}$ reduction in $PM_{2.5}$ on a cohort of people: all the people alive in 2008. Now let us look at the more realistic effects. People are born every year and are thus "new entrants" that should be included in our calculations. We must take these into account: they, too, gain by living in a country in which $PM_{2.5}$ has been reduced by $1\ \mu g\ m^{-3}$. Panel (a) (solid line) shows that if we include the new entrants, the gain in life years per year reaches a maximum in about 2050 and thereafter remains constant. Panel (b) (solid line) shows that the cumulative years of life lived continues to rise, from 2008 to 2050 and on to 2113 and on beyond 2113. In fact,

---

Figure 15.1 as annual (a) and cumulative (b) gains in numbers of life-years, *(continued)* and as annual reductions in numbers of deaths (c). Dashed line shows life years gained and reduction in deaths per year in the current population; solid line life-years gained and reduction in deaths per year including new births from 2009. Reproduced from ref. 4, https://assets.publishing.service.gov.uk/government/uploads/system/uploads/attachment_data/file/304641/COMEAP_mortality_effects_of_long_term_exposure.pdf, under the terms of the Open Government Licence v3.0, https://www.nationalarchives.gov.uk/doc/open-government-licence/version/3/.

it continues to rise forever! The benefits of the 1 µg m^{-3} reduction in PM$_{2.5}$ in 2008 continue to accrue: those born in 2008 will benefit, those born in 2108 will benefit, those born in 2208 will benefit and so will all those born after 2208 and so on.

Now we have a meaningful way of representing the benefits to public health of reducing concentrations of fine particles (PM$_{2.5}$). If we assume linearity of effects then we may extrapolate directly from the numbers given above to the effects of larger, or smaller, reductions in ambient concentrations of fine particles (PM$_{2.5}$).

15.4   Further Consideration of the Uses of "Attributable Deaths", "Years of Life Gained" and "Average Loss of Life Expectancy"

We have seen that the coefficient reported[1,2] may be used to estimate the *reduction* in the number of attributable deaths and also the years of life *gained* by *reductions* in the level of fine particles (PM$_{2.5}$). These two calculations are merely two rather different ways of looking at the same thing: the impact on public health of particulate air pollution.

Calculation of attributable deaths has proved useful in drawing attention of the public to the problem posed by air pollution. As we have seen, however, the estimate is open to misinterpretation. The most likely misinterpretation is that 29 000 people die, in the UK, each year as a direct result of exposure to fine particles (PM$_{2.5}$). This misinterpretation is not the fault of the public but, rather, of the scientists who chose the word "attributable". In everyday usage the word attributable implies a complete explanation of an event: my car coming to a stop is attributable to there being no more petrol in the tank; the failure of the tomatoes to ripen is attributable to a lack of sunlight. It is, then, hardly surprising that when told that 29 000 attributable deaths are associated with a type of air pollution that the public infer that 29 000 deaths are explained completely by that type of air pollution. The statement seems very similar to: 40 000 deaths are caused each year by lung cancer.

The essential difference between the statements is due to the difference between the methods used to discover the size of the effects. In the case of deaths from lung cancer the information comes from death certificates: people dying from lung cancer are recorded as dying from lung cancer and the total number of deaths due to lung cancer is arrived at by the simple process of counting. In the case of attributable deaths associated with air pollution there is no information available from death certificates: no doctor puts "air pollution" as the cause of death on a death certificate. On the contrary, the critical information is derived from epidemiological studies of the relative risk of death in areas with "high" as compared with areas with "low" levels of air pollution.

The reader is now left with a choice. Which of the following statements best conveys the effects of fine particulate air pollution on health in the UK?

1. Current levels of fine particulate air pollution ($PM_{2.5}$) are associated with 29 000 attributable deaths per year.
2. Current levels of fine particulate air pollution ($PM_{2.5}$) are associated with an average loss of life expectancy across the population of 3 days.
3. Current levels of fine particulate air pollution ($PM_{2.5}$) are associated with a loss of total population life expectancy of 340 000 years.
4. For a $1 \ \mu g \ m^{-3}$ reduction in $PM_{2.5}$ about 4 000 000 life years could be added to the total life years of the population over the next 106 years with an average increase in life expectancy of about 20 days.

Each statement is arithmetically correct, and each statement tells us something about the effects of fine particulate air pollution ($PM_{2.5}$) on public health in the UK. If one wished to draw attention to the problem one might well choose the first statement; if one wished to estimate the benefits of reducing levels of fine particulate air pollution one would certainly choose the last.

15.4.1 Conclusion Regarding Effects of Long-term Exposure to Air Pollutants

Quantification of the effects of long-term exposure air pollutants on public health is possible if coefficients reflecting the relative risk of death across a range of long-term average concentrations of air pollutants are available. The results of such calculations can be expressed in a variety of ways but these differ only in terms of the arithmetic processes involved in their calculation. The various results have differing impacts on the public mind and vary in their usefulness to those charged with estimating the benefits to public health of reductions in levels of air pollution. For the moment we have reliable data relating to the effects of long-term exposure to fine particulate air pollutants monitored as $PM_{2.5}$; similar information relating to ozone and nitrogen dioxide may become available.

15.5 Quantification of the Effects of Short-term Exposure to Air Pollutants

This section will be shorter than the above. Time series studies reveal association between daily concentrations of air pollutants (usually expressed as 24 hour average concentrations) and effects on health indicated by daily deaths, admissions to hospital, consultations with general practitioners and several other indices. Here we shall take just one example.

Let us accept that a coefficient linking the 24 hour concentration of particulate matter monitored as PM_{10} and the daily number of all cause, non-accidental, deaths can be expressed as an increase in deaths of 0.6% per 10 $\mu g\, m^{-3}$ PM_{10}. The 24 hour average concentration will of course vary from day to day, not as a result of variation in the strength of sources but, rather, because of variation in the weather. Let us suppose we average the 24 hour average concentrations across a year. As was suggested above an estimate of the effect on deaths can be easily made. Let the total number of deaths per year be 500 000. Then, let us say, by way of an example, that the annual average concentration of particulate matter (PM_{10}) is 20 $\mu g\, m^{-3}$, then, the excess deaths, attributable to particulate matter (PM_{10}) will be about 1.2% of 500 000, *i.e.* about 6000. Coefficients derived from time series studies are available for nitrogen dioxide, sulphur dioxide and ozone. Of course, as in the case of the effects of long-term exposure to air pollutants, we might have concerns about the independence of the effects implied by such coefficients. In the case of ozone we might also be concerned about the possibility of a threshold of effect and choose to limit the calculations to the summer months when ozone concentrations are highest and effects most likely.

It will have been noticed that the estimated effect of daily variations of concentrations of particulate matter monitored as PM_{10}, for that is what time series studies tell us about, are much lower than that of long-term exposure to particulate matter monitored as $PM_{2.5}$. Given that the calculations are done in the same way and that the respective coefficients differ by an order of magnitude, this is to be expected. What is more interesting is the question of the extent to which the calculations overlap one another. This question has not yet been settled but most authorities advise against a simple adding of the calculated effects. In practice, estimates of impacts based on time series studies reflecting the effects of short-term variations in concentrations of pollutants have been overtaken by estimates of the effects of long-term exposure.

References

1. C. A. Pope III, M. J. Thun, M. M. Namboodiri, D. W. Dockery, J. S. Evans, F. E. Speizer and C. W. Heath, Particulate air pollution as a predictor of mortality in a prospective cohort study of U S adults, *Am. J. Respir. Crit. Care Med.*, 1995, **151**, 669–674.
2. C. A. Pope III, R. T. Burnett, M. J. Thun, E. E. Calle, D. Krewski, K. Ito and G. D. Thurston, Lung cancer, cardiopulmonary mortality and long-term exposure to fine particulate air pollution, *JAMA, J. Am. Med. Assoc.*, 2002, **287**, 1132–1141.
3. A. J. Cohen, M. Brauer, R. Burnett and A. H. Ross, *et al.*, Estimates and 25-year trends of the global burden of disease attributable to ambient air pollution: an analysis of data from the Global Burden of Disease Study 2015, *Lancet*, 2017, **389**, 1907–1918.
4. Department of Health and Committee on the Medical Effects of Air Pollutants, *The Mortality Effects of Long-term Exposure to Particulate Air Pollution in the United Kingdom*, 98 pages, 2010.

Appendix

A.1   Area Under the Normal Curve

Table A1 gives the area under a normal curve between value z and the mean, for a standard normal distribution where *mean = 0* and *standard deviation = 1*.

Find your z value by reading down the left column for the units and tenths and along the top row for the hundredths (Figure A1).

A.2   Ordinates (y Values) for the Normal Probability Curve

Table A2 gives the values of the ordinates of the standard normal distribution curve (y, see eqn (6.4)) for specified values of z.

Find your z value by reading down the left column for the units and tenths and along the top row for the hundredths (Figure A2).

A.3   Chi-squared Distribution

Table A3 gives the values of χ^2 for a given area (p) of the right tail of the chi-squared distribution (Figure A3) for various degrees of freedom (k).

Basic Mathematics for Students of Air Pollutants
By Robert Maynard and Richard Atkinson
© Robert Maynard and Richard Atkinson 2024
Published by the Royal Society of Chemistry, www.rsc.org

Table A1 The area below a standard normal distribution. Reproduced from ref. 1 with permission from John Wiley & Sons, Copyright © 2007 John Wiley & Sons, Inc.

z	0	0.01	0.02	0.03	0.04	0.05	0.06	0.07	0.08	0.09
0	0	0.004	0.008	0.012	0.016	0.0199	0.0239	0.0279	0.0319	0.0359
0.1	0.0398	0.0438	0.0478	0.0517	0.0557	0.0596	0.0636	0.0675	0.0714	0.0753
0.2	0.0793	0.0832	0.0871	0.091	0.0948	0.0987	0.1026	0.1064	0.1103	0.1141
0.3	0.1179	0.1217	0.1255	0.1293	0.1331	0.1368	0.1406	0.1443	0.148	0.1517
0.4	0.1554	0.1591	0.1628	0.1664	0.17	0.1736	0.1772	0.1808	0.1844	0.1879
0.5	0.1915	0.195	0.1985	0.2019	0.2054	0.2088	0.2123	0.2157	0.219	0.2224
0.6	0.2257	0.2291	0.2324	0.2357	0.2389	0.2422	0.2454	0.2486	0.2517	0.2549
0.7	0.258	0.2611	0.2642	0.2673	0.2703	0.2734	0.2764	0.2794	0.2823	0.2852
0.8	0.2881	0.291	0.2939	0.2967	0.2995	0.3023	0.3051	0.3078	0.3106	0.3133
0.9	0.3159	0.3186	0.3212	0.3238	0.3264	0.3289	0.3315	0.334	0.3365	0.3389
1	0.3413	0.3438	0.3461	0.3485	0.3508	0.3531	0.3554	0.3577	0.3599	0.3621
1.1	0.3643	0.3665	0.3686	0.3708	0.3729	0.3749	0.377	0.379	0.381	0.383
1.2	0.3849	0.3869	0.3888	0.3907	0.3925	0.3944	0.3962	0.398	0.3997	0.4015
1.3	0.4032	0.4049	0.4066	0.4082	0.4099	0.4115	0.4131	0.4147	0.4162	0.4177
1.4	0.4192	0.4207	0.4222	0.4236	0.4251	0.4265	0.4279	0.4292	0.4306	0.4319
1.5	0.4332	0.4345	0.4357	0.437	0.4382	0.4394	0.4406	0.4418	0.4429	0.4441
1.6	0.4452	0.4463	0.4474	0.4484	0.4495	0.4505	0.4515	0.4525	0.4535	0.4545
1.7	0.4554	0.4564	0.4573	0.4582	0.4591	0.4599	0.4608	0.4616	0.4625	0.4633
1.8	0.4641	0.4649	0.4656	0.4664	0.4671	0.4678	0.4686	0.4693	0.4699	0.4706
1.9	0.4713	0.4719	0.4726	0.4732	0.4738	0.4744	0.475	0.4756	0.4761	0.4767
2	0.4772	0.4778	0.4783	0.4788	0.4793	0.4798	0.4803	0.4808	0.4812	0.4817
2.1	0.4821	0.4826	0.483	0.4834	0.4838	0.4842	0.4846	0.485	0.4854	0.4857
2.2	0.4861	0.4864	0.4868	0.4871	0.4875	0.4878	0.4881	0.4884	0.4887	0.489
2.3	0.4893	0.4896	0.4898	0.4901	0.4904	0.4906	0.4909	0.4911	0.4913	0.4916
2.4	0.4918	0.492	0.4922	0.4925	0.4927	0.4929	0.4931	0.4932	0.4934	0.4936
2.5	0.4938	0.494	0.4941	0.4943	0.4945	0.4946	0.4948	0.4949	0.4951	0.4952
2.6	0.4953	0.4955	0.4956	0.4957	0.4959	0.496	0.4961	0.4962	0.4963	0.4964
2.7	0.4965	0.4966	0.4967	0.4968	0.4969	0.497	0.4971	0.4972	0.4973	0.4974
2.8	0.4974	0.4975	0.4976	0.4977	0.4977	0.4978	0.4979	0.4979	0.498	0.4981
2.9	0.4981	0.4982	0.4982	0.4983	0.4984	0.4984	0.4985	0.4985	0.4986	0.4986
3	0.4987	0.4987	0.4987	0.4988	0.4988	0.4989	0.4989	0.4989	0.499	0.499
3.1	0.499	0.4991	0.4991	0.4991	0.4992	0.4992	0.4992	0.4992	0.4993	0.4993
3.2	0.4993	0.4993	0.4994	0.4994	0.4994	0.4994	0.4994	0.4995	0.4995	0.4995
3.3	0.4995	0.4995	0.4995	0.4996	0.4996	0.4996	0.4996	0.4996	0.4996	0.4997
3.4	0.4997	0.4997	0.4997	0.4997	0.4997	0.4997	0.4997	0.4997	0.4997	0.4998
3.5	0.4998	0.4998	0.4998	0.4998	0.4998	0.4998	0.4998	0.4998	0.4998	0.4998
3.6	0.4998	0.4998	0.4999	0.4999	0.4999	0.4999	0.4999	0.4999	0.4999	0.4999
3.8	0.4999	0.4999	0.4999	0.4999	0.4999	0.4999	0.4999	0.4999	0.4999	0.4999
3.9	0.4999	0.4999	0.4999	0.4999	0.4999	0.4999	0.4999	0.4999	0.4999	0.4999

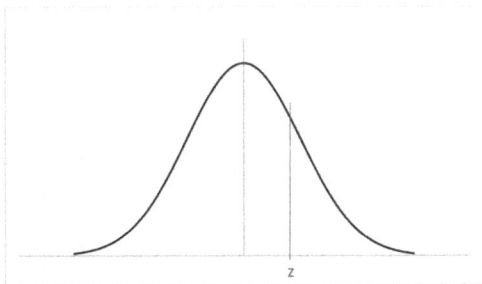

Figure A1 Normal distribution with *z* marked.

Table A2 The *y* ordinate for a standard normal distribution. Reproduced from ref. 1 with permission from John Wiley & Sons, Copyright © 2007 John Wiley & Sons, Inc.

z	0	0.01	0.02	0.03	0.04	0.05	0.06	0.07	0.08	0.09
0	0.3989	0.3989	0.3989	0.3988	0.3986	0.3984	0.3982	0.398	0.3977	0.3973
0.1	0.397	0.3965	0.3961	0.3956	0.3951	0.3945	0.3939	0.3932	0.3925	0.3918
0.2	0.391	0.3902	0.3894	0.3885	0.3876	0.3867	0.3857	0.3847	0.3836	0.3825
0.3	0.3814	0.3802	0.379	0.3778	0.3765	0.3752	0.3739	0.3725	0.3712	0.3697
0.4	0.3683	0.3668	0.3653	0.3637	0.3621	0.3605	0.3589	0.3572	0.3555	0.3538
0.5	0.3521	0.3503	0.3485	0.3467	0.3448	0.3429	0.341	0.3391	0.3372	0.3352
0.6	0.3332	0.3312	0.3292	0.3271	0.3251	0.323	0.3209	0.3187	0.3166	0.3144
0.7	0.3123	0.3101	0.3079	0.3056	0.3034	0.3011	0.2989	0.2966	0.2943	0.292
0.8	0.2897	0.2874	0.285	0.2827	0.2803	0.278	0.2756	0.2732	0.2709	0.2685
0.9	0.2661	0.2637	0.2613	0.2589	0.2565	0.2541	0.2516	0.2492	0.2468	0.2444
1	0.242	0.2396	0.2371	0.2347	0.2323	0.2299	0.2275	0.2251	0.2227	0.2203
1.1	0.2179	0.2155	0.2131	0.2107	0.2083	0.2059	0.2036	0.2012	0.1989	0.1965
1.2	0.1942	0.1919	0.1895	0.1872	0.1849	0.1826	0.1804	0.1781	0.1758	0.1736
1.3	0.1714	0.1691	0.1669	0.1647	0.1626	0.1604	0.1582	0.1561	0.1539	0.1518
1.4	0.1497	0.1476	0.1456	0.1435	0.1415	0.1394	0.1374	0.1354	0.1334	0.1315
1.5	0.1295	0.1276	0.1257	0.1238	0.1219	0.12	0.1182	0.1163	0.1145	0.1127
1.6	0.1109	0.1092	0.1074	0.1057	0.104	0.1023	0.1006	0.0989	0.0973	0.0957
1.7	0.094	0.0925	0.0909	0.0893	0.0878	0.0863	0.0848	0.0833	0.0818	0.0804
1.8	0.079	0.0775	0.0761	0.0748	0.0734	0.0721	0.0707	0.0694	0.0681	0.0669
1.9	0.0656	0.0644	0.0632	0.062	0.0608	0.0596	0.0584	0.0573	0.0562	0.0551
2	0.054	0.0529	0.0519	0.0508	0.0498	0.0488	0.0478	0.0468	0.0459	0.0449
2.1	0.044	0.0431	0.0422	0.0413	0.0404	0.0396	0.0387	0.0379	0.0371	0.0363
2.2	0.0355	0.0347	0.0339	0.0332	0.0325	0.0317	0.031	0.0303	0.0297	0.029
2.3	0.0283	0.0277	0.027	0.0264	0.0258	0.0252	0.0246	0.0241	0.0235	0.0229
2.4	0.0224	0.0219	0.0213	0.0208	0.0203	0.0198	0.0194	0.0189	0.0184	0.018
2.5	0.0175	0.0171	0.0167	0.0163	0.0158	0.0154	0.0151	0.0147	0.0143	0.0139

(*continued*)

Table A2 (*Continued*)

z	0	0.01	0.02	0.03	0.04	0.05	0.06	0.07	0.08	0.09
2.6	0.0136	0.0132	0.0129	0.0126	0.0122	0.0119	0.0116	0.0113	0.011	0.0107
2.7	0.0104	0.0101	0.0099	0.0096	0.0093	0.0091	0.0088	0.0086	0.0084	0.0081
2.8	0.0079	0.0077	0.0075	0.0073	0.0071	0.0069	0.0067	0.0065	0.0063	0.0061
2.9	0.006	0.0058	0.0056	0.0055	0.0053	0.0051	0.005	0.0048	0.0047	0.0046
3	0.0044	0.0043	0.0042	0.004	0.0039	0.0038	0.0037	0.0036	0.0035	0.0034
3.1	0.0033	0.0032	0.0031	0.003	0.0029	0.0028	0.0027	0.0026	0.0025	0.0025
3.2	0.0024	0.0023	0.0022	0.0022	0.0021	0.002	0.002	0.0019	0.0018	0.0018
3.3	0.0017	0.0017	0.0016	0.0016	0.0015	0.0015	0.0014	0.0014	0.0013	0.0013
3.4	0.0012	0.0012	0.0012	0.0011	0.0011	0.001	0.001	0.001	0.0009	0.0009
3.5	0.0009	0.0008	0.0008	0.0008	0.0008	0.0007	0.0007	0.0007	0.0007	0.0006
3.6	0.0006	0.0006	0.0006	0.0005	0.0005	0.0005	0.0005	0.0005	0.0005	0.0004
3.7	0.0004	0.0004	0.0004	0.0004	0.0004	0.0004	0.0003	0.0003	0.0003	0.0003
3.8	0.0003	0.0003	0.0003	0.0003	0.0003	0.0002	0.0002	0.0002	0.0002	0.0002
3.9	0.0002	0.0002	0.0002	0.0002	0.0002	0.0002	0.0002	0.0002	0.0001	0.0001

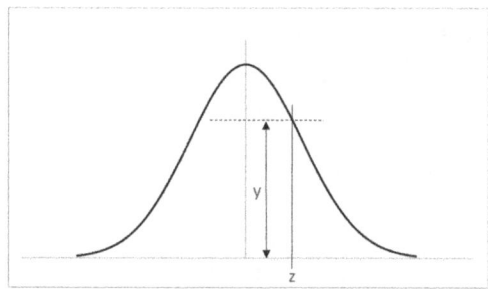

Figure A2 Normal distribution indicating height y at point z.

Table A3 Table of values of χ^2 in a chi-squared distribution with k degrees of freedom such that p is the area between χ^2 and $+\infty$. Reproduced from ref. 2, https://en.wikibooks.org/wiki/Engineering_Tables/Chi-Squared_Distribution, under the terms of the CC BY-SA 4.0 license https://creativecommons.org/licenses/by-sa/4.0/.

k	Probability content, p, between χ^2 and $+\infty$														
	0.995	0.99	0.975	0.95	0.9	0.75	0.5	0.25	0.1	0.05	0.025	0.01	0.005	0.002	0.001
1	3.93×10^{-5}	1.57×10^{-4}	9.82×10^{-4}	0.00393	0.0157	0.102	0.455	1.323	2.706	3.841	5.024	6.635	7.879	9.55	10.828
2	0.01	0.0201	0.0506	0.103	0.211	0.575	1.386	2.773	4.605	5.991	7.378	9.21	10.597	12.429	13.816
3	0.0717	0.115	0.216	0.352	0.584	1.213	2.366	4.108	6.251	7.815	9.348	11.345	12.838	14.796	16.266
4	0.207	0.297	0.484	0.711	1.064	1.923	3.357	5.385	7.779	9.488	11.143	13.277	14.86	16.924	18.467
5	0.412	0.554	0.831	1.145	1.61	2.675	4.351	6.626	9.236	11.07	12.833	15.086	16.75	18.907	20.515
6	0.676	0.872	1.237	1.635	2.204	3.455	5.348	7.841	10.645	12.592	14.449	16.812	18.548	20.791	22.458
7	0.989	1.239	1.69	2.167	2.833	4.255	6.346	9.037	12.017	14.067	16.013	18.475	20.278	22.601	24.322
8	1.344	1.646	2.18	2.733	3.49	5.071	7.344	10.219	13.362	15.507	17.535	20.09	21.955	24.352	26.124
9	1.735	2.088	2.7	3.325	4.168	5.899	8.343	11.389	14.684	16.919	19.023	21.666	23.589	26.056	27.877
10	2.156	2.558	3.247	3.94	4.865	6.737	9.342	12.549	15.987	18.307	20.483	23.209	25.188	27.722	29.588
11	2.603	3.053	3.816	4.575	5.578	7.584	10.341	13.701	17.275	19.675	21.92	24.725	26.757	29.354	31.264
12	3.074	3.571	4.404	5.226	6.304	8.438	11.34	14.845	18.549	21.026	23.337	26.217	28.3	30.957	32.909
13	3.565	4.107	5.009	5.892	7.042	9.299	12.34	15.984	19.812	22.362	24.736	27.688	29.819	32.535	34.528
14	4.075	4.66	5.629	6.571	7.79	10.165	13.339	17.117	21.064	23.685	26.119	29.141	31.319	34.091	36.123
15	4.601	5.229	6.262	7.261	8.547	11.037	14.339	18.245	22.307	24.996	27.488	30.578	32.801	35.628	37.697
16	5.142	5.812	6.908	7.962	9.312	11.912	15.338	19.369	23.542	26.296	28.845	32	34.267	37.146	39.252
17	5.697	6.408	7.564	8.672	10.085	12.792	16.338	20.489	24.769	27.587	30.191	33.409	35.718	38.648	40.79
18	6.265	7.015	8.231	9.39	10.865	13.675	17.338	21.605	25.989	28.869	31.526	34.805	37.156	40.136	42.312
19	6.844	7.633	8.907	10.117	11.651	14.562	18.338	22.718	27.204	30.144	32.852	36.191	38.582	41.61	43.82
20	7.434	8.26	9.591	10.851	12.443	15.452	19.337	23.828	28.412	31.41	34.17	37.566	39.997	43.072	45.315
21	8.034	8.897	10.283	11.591	13.24	16.344	20.337	24.935	29.615	32.671	35.479	38.932	41.401	44.522	46.797
22	8.643	9.542	10.982	12.338	14.041	17.24	21.337	26.039	30.813	33.924	36.781	40.289	42.796	45.962	48.268
23	9.26	10.196	11.689	13.091	14.848	18.137	22.337	27.141	32.007	35.172	38.076	41.638	44.181	47.391	49.728
24	9.886	10.856	12.401	13.848	15.659	19.037	23.337	28.241	33.196	36.415	39.364	42.98	45.559	48.812	51.179
25	10.52	11.524	13.12	14.611	16.473	19.939	24.337	29.339	34.382	37.652	40.646	44.314	46.928	50.223	52.62

(continued)

Table A3 (*Continued*)

k	Probability content, p, between χ^2 and $+\infty$														
	0.995	0.99	0.975	0.95	0.9	0.75	0.5	0.25	0.1	0.05	0.025	0.01	0.005	0.002	0.001
26	11.16	12.198	13.844	15.379	17.292	20.843	25.336	30.435	35.563	38.885	41.923	45.642	48.29	51.627	54.052
27	11.808	12.879	14.573	16.151	18.114	21.749	26.336	31.528	36.741	40.113	43.195	46.963	49.645	53.023	55.476
28	12.461	13.565	15.308	16.928	18.939	22.657	27.336	32.62	37.916	41.337	44.461	48.278	50.993	54.411	56.892
29	13.121	14.256	16.047	17.708	19.768	23.567	28.336	33.711	39.087	42.557	45.722	49.588	52.336	55.792	58.301
30	13.787	14.953	16.791	18.493	20.599	24.478	29.336	34.8	40.256	43.773	46.979	50.892	53.672	57.167	59.703

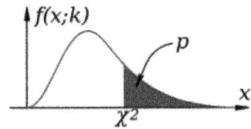

Figure A3 Chi squared distribution. Reproduced from ref. 2, https://en.wiki-books.org/wiki/Engineering_Tables/Chi-Squared_Distribution, under the terms of the CC BY-SA 4.0 license https://creative-commons.org/licenses/by-sa/4.0/.

References

1. W. M. Bolstad, Appendix B: Use of Statistical Tables, in *Introduction to Bayesian Statistics*, ed. W. M. Bolstad, 2007, DOI: 10.1002/9780470181188.app2.
2. Wikimedia Foundation, Engineering Tables/Chi-Squared Distribution, *Wikibooks* [updated 5 August 2023, cited 7 September 2023], available from https://en.wiki-books.org/wiki/Engineering_Tables/Chi-Squared_Distribution

Further Reading

Mathematics

As noted in the preface we have not included any textbooks of higher mathematics in this section. This is not, of course, to say that the reader would not be better equipped if he or she did have a knowledge of higher mathematics. What is higher mathematics? Most people would agree that the inclusion of differential and integral calculus is the hallmark of the subject; neither is required for the work outlined in this book. One of us (RM) has found that older textbooks of school algebra provide much of what is needed and we have listed a few such books. We have already mentioned, in the preface, "*Mathematics for Biologists*" by Crowe and Crowe. This is an outstanding book. It contains all the mathematics needed for the topics we have considered *and* an introduction to the calculus.

C. V. Durell, G. W. Palmer and R. M. Wright, *Elementary Algebra*, G Bell and Sons Ltd, London, 1921. More recent editions are available.

W. G. Borchardt, *Elementary Algebra*, Rivingtons, London, 6th edn, 1914.

Both these books are obviously elderly. But they provide admirable introductions to algebra and contain more than is found in some modern introductory school textbooks.

Alan and A. Crowe, *Mathematics for Biologists*, Academic Press, London and New York, 1969.

Basic Mathematics for Students of Air Pollutants
By Robert Maynard and Richard Atkinson
© Robert Maynard and Richard Atkinson 2024
Published by the Royal Society of Chemistry, www.rsc.org

Statistics

Much the same as was said above could be said of textbooks of statistics: the advanced books include use of the calculus, the elementary accounts do not. We have, however, included one book on mathematical statistics, by Hoel. The reader will note that we have also included Hoel's *"Elementary Statistics"*: if the reader uses that then he or she might go on to look at Hoel's more advanced book.

Books Not Involving Use of Calculus

Here the choice is very wide: we have used the following books and found them helpful:

H. L. Alder and E. B. Roessler, *Introduction to Probability and Statistics*, W H Freeman and Company, San Francisco, 5th edn, 1972. More recent editions are available. A very clear account including a few non-parametric methods.

P. G. Hoel, *Elementary Statistics*, John Wiley and Sons. Inc., New York, 2nd edn, 1966. Elegant development of the theory.

G. U. Yule and M. G. Kendall, *Introduction to the Theory of Statistics*, Griffin, London, 14th edn, 1950. This is rightly regarded as a classic. The many examples, an unusually clear style and the occasional jokes are very appealing.

G. W. Snedecor and W. G. Cochran, *Statistical Methods*, Iowa State University Press, Ames, Iowa, USA, 6th edn, 1967. A more recent edition is available. This is another classic handbook.

S. Siegel, *Nonparametric Statistics for the Behavioural Sciences*, McGraw-Hill Kogakusha Ltd, Tokyo, 1956. This is the book that introduced most research workers in the biological sciences to non-parametric statistical methods.

Scientific Tables. (Documenta Geigy), ed. K. Diem and C. Lentner, Geigy Pharmaceuticals, Macclesfield, UK, 7th edn, 1970. An invaluable book containing a superb collection of statistical tables and an abbreviated, but excellent, account of statistical methods. The seventh edition was a one volume work; the eighth is in six volumes with the second devoted to an *Introduction to Statistics, Statistical Tables and Mathematical Formulae*.

Mathematical Statistics, Involving Use of Calculus

P. G. Hoel, *Introduction to Mathematical Statistics*, John Wiley and Sons, Inc., New York, 3rd edn, 1954. More recent editions are available.

Other Works on Statistics

M. G. Kendall and W. R. Buckland, *A Dictionary of Statistical Terms*, Oliver and Boyd, Edinburgh, 1960.

M. Bland, *An Introduction to Medical Statistics*, Oxford University Press, 2015, Biometry – 446 pages.

P. Armitage, G. Berry and J. N. S. Matthews, *Statistical Methods in Medical Research*, John Wiley & Sons, 2008.

Aerosols

Here the choice is easy: Hinds' *Aerosol Technology* is the essential book for all aerosol scientists.

W. C. Hinds, *Aerosol Technology*, John Wiley and Sons Inc., New York, 2nd edn, 1999.

G. Herdan, *Small Particle Statistics*, Butterworths, London, 2nd edn, 1960. This is an advanced work which contains much of the statistical theory required by aerosol scientists.

Subject Index

Page references that are solely to a figure or table are indicated with *italics* or a suffix T respectively. Other references may include figures or tables as well as text.

土49